A. Böttcher · B. Silbermann

Invertibility and Asymptotics of Toeplitz Matrices

Mathematical Research · Mathematische Forschung

Wissenschaftliche Beiträge
herausgegeben von der
Akademie der Wissenschaften der DDR
Institut für Mathematik

Band 17

Invertibility and Asymptotics
of Toeplitz Matrices

by

A. Böttcher
B. Silbermann

Invertibility and Asymptotics of Toeplitz Matrices

by
Albrecht Böttcher
and
Bernd Silbermann

Akademie-Verlag · Berlin 1983

Verfasser:
Dipl.-Math. Albrecht Böttcher
Prof. Dr. sc. Bernd Silbermann
Technische Hochschule Karl-Marx-Stadt
Sektion Mathematik
DDR-9010 Karl-Marx-Stadt
PSF 964

Die Titel dieser Schriftenreihe werden vom Originalmanuskript der Autoren reproduziert

Erschienen im Akademie-Verlag, DDR-1086 Berlin, Leipziger Str. 3-4

Lizenznummer: 202·100/412/83
Printed in the German Democratic Republic
Gesamtherstellung: VEB Kongreß- und Werbedruck, 9273 Oberlungwitz
Lektor: Dr. Reinhard Höppner
Umschlag: Karl Salzbrunn
LSV 1075
Bestellnummer: 763 282 2 (2182/17)
DDR 25,- M
ISSN 0138 - 3019

PREFACE

The study of infinite and finite Toeplitz operators and matrices has been an active field of research for a long time and remains so to date. The reasons are of a quite different nature. On the one hand, besides self-adjoint or unitary operators, the Toeplitz operators belong to the few non-trivial classes of operators about whose structure one has a rather detailed knowledge. Moreover, the development of the last twenty years has brought to light manifold and fertile connections between Toeplitz operators and several other areas in mathematics, and in many cases one has not fully understood and exhausted these connections yet.

On the other hand, Toeplitz operator occur in a variety of applications. The physicist, for instance, has been concerned with Toeplitz matrices and determinants for a long time and many problems were first posed in the physics literature. A fruitful (and often quoted) example is the two-dimensional Ising model, leading in a natural way to the study of large Toeplitz determinants with singular generating functions (see Fisher, Hartwig [1], Montroll, Potts, Ward [1], Wu [1], Lenard [1], or McCoy, Wu [1]). Toeplitz determinants arise from some other questions of statistical physics, too. Examples are the dimer and monomer problems (see Fisher, Hartwig [1], Fisher [1], Fisher, Stephenson [1], or Hartwig [1]) or the one-dimensional Bose gas with delta function interactions (see Lenard [2]). As a rule, in all these examples, the order of the Toeplitz determinant is equal to the number of particles under consideration, that is, about 10^{23}. Clearly, such determinants can only be computed asymptotically. Other fields offering advantageous applications of Toeplitz matrices and determinants are: orthogonal polynomials, random walks, prediction theory, stationary Gaussian processes (see Kac [1], Grenander, Szegö [1], Helson, Szegö [1], Fisher, Hartwig [1]. Peller, Hruščev [1], [2]), or information and control theory (see Mullis, Roberts [1], [2], Kailath, Kung, Morf [1]).

The study of the asymptotic behavior of Toeplitz determinants has been both origin and destination of our investigations. We emphasize the view that the natural approach to this set of problems rests on the finite section method, the study of which, on its part, requires precise knowledge about infinite Toeplitz and Hankel matrices (operators).

However, there are many questions arising quite naturally in the context of Toeplitz determinants, but which are of independent interest and play an important part in connection with several other problems, too. This concerns above all the finite section method itself. Such topics are also algebras related to operator ideals or the Wiener-Hopf factorization. Therefore we have paid due attention to those subjects and dwelt on them longer than necessary for their application to Toeplitz determinants.

Thus, the material of the book is arranged as follows. The first chapter is a collection of some (in many cases standard) facts from functional analysis. The second chapter is devoted to the general theory of Toeplitz and Hankel operators, on the basis of which an almost self-contained theory of the finite section method for Toeplitz operators is then founded in the third chapter. We propose two approaches to the finite section method, which we have named "constructive" and "local". The local approach consists in a reformulation of the problem in the language of Banach algebras and a subsequent reduction of the problem to one for a (much simpler) "local representative" by means of localization techniques. The corresponding local representatives can then be studied with the methods of the constructive theory. The following two chapters present some results interesting by themselves, but mainly dictated by the needs of the sixth chapter, where a comprehensive theory of Toeplitz determinants (Szegö's limit theorem, its refinements and generalizations) for elliptic symbols is established. This includes estimates for the higher-order correction terms, generating functions with weakened smoothness or even with jumps, cases where Szegö's asymptotic formula becomes an equality, rational generating functions and the formula of Day, as well as Devinatz's converse of Szegö's limit theorem for positive-definite generating functions. As far as we know, there is up to now no treatise dealing with all these aspects from a unique point of view. Finally, the seventh chapter contains a survey of the present state for the non-elliptic case and some new results about Toeplitz determinants with singular generating functions.

At the end of each chapter source notes and comments follow. The sources cited reflect our knowledge and the literature available to us. Therefore omissions certainly occur. In no case does the absence of a reference imply any claim to originality on our part. In addition, at the end of the third and sixth chapter some information about related topics and further problems is given, which for lack of space could not be included in the text.

Much of the material presented in this volume is well-known to specialists, some results were obtained by the authors in recent years and are published here for the first time. We hope the book can serve the beginner as an introduction to the subject, and we would be glad if the specialist could find something that is new to him of if this work contained a few facts of interest for anybody working in applied research.

We would like to thank our colleagues S. Dümmel, G. Heinig, K. Rost, S. Roch, and J. Schulz for many fruitful conversations about various topics in this book and for reading large parts of the manuscript. Finally, we are extremely grateful to Mrs. M. Graupner for her patient excellent work in typing the whole manuscript.

C O N T E N T S

CHAPTER 1

SOME FACTS FROM FUNCTIONAL ANALYSIS

OPERATOR IDEALS

1.1. Bounded and compact operators. Let X and Y be Banach spaces. We denote by $\mathfrak{B}(X,Y)$ the linear space of all (bounded and linear) operators of X into Y. Furthermore, let $\mathfrak{C}_\infty(X,Y)$ denote the collection of all compact operators of $\mathfrak{B}(X,Y)$ and let $\mathfrak{C}_0(X,Y)$ be the set of all finite-rank operators, i.e. of all operators $F \in \mathfrak{B}(X,Y)$ for which $\dim F(X) < \infty$.
In the case $X = Y$ we shall write $\mathfrak{B}(X) = \mathfrak{B}(X,X)$, $\mathfrak{C}_\infty(X) = \mathfrak{C}_\infty(X,X)$, $\mathfrak{C}_0(X) = \mathfrak{C}_0(X,X)$.
A sequence $\{A_n\}$ of operators $A_n \in \mathfrak{B}(X,Y)$ is said to converge to an operator $A \in \mathfrak{B}(X,Y)$

(a) weakly, if $f(A_n x) - f(Ax) \to 0$ for each $f \in Y^*$ and each $x \in X$;

(b) strongly, if $\|A_n x - Ax\|_Y \to 0$ for each $x \in X$;

(c) uniformly, if $\|A_n - A\| \to 0$, where for $B \in \mathfrak{B}(X,Y)$

$$\|B\| := \sup_{\|x\| \leq 1} \|Bx\|. \tag{1}$$

Compact operators have the following important property:

(d) If $A_n \to A$ weakly and $K \in \mathfrak{C}_\infty(Y,Z)$, then $KA_n \to KA$ strongly.

Under the operator norm (1) the linear set $\mathfrak{B}(X,Y)$ becomes a Banach space and $\mathfrak{B}(X)$ a Banach algebra. Then $\mathfrak{C}_\infty(X,Y)$ is a closed subspace of $\mathfrak{B}(X,Y)$. In general the subspace $\mathfrak{C}_0(X,Y) \subset \mathfrak{B}(X,Y)$ is not closed, but its closure is contained in $\mathfrak{C}_\infty(X,Y)$. The following implications hold:

$$A\in\mathfrak{B}(X,Y),\ K\in\mathfrak{C}_\infty(Y,Z),\ B\in\mathfrak{B}(Z,V) \implies BKA \in \mathfrak{C}_\infty(X,V),$$

$$A\in\mathfrak{B}(X,Y),\ F\in\mathfrak{C}_0(Y,Z),\ B\in\mathfrak{B}(Z,V) \implies BFA \in \mathfrak{C}_0(X,V).$$

In particular, $\mathfrak{C}_\infty(X)$ is a closed two-sided ideal in $\mathfrak{B}(X)$, and $\mathfrak{C}_0(X)$ is a two-sided, but in general not closed, ideal in $\mathfrak{B}(X)$.

In the special case where X and Y are separable infinite-dimensional Hilbert spaces, we shall write $X = H_1$, $Y = H_2$ and for $X = Y$ simply $X = Y = H$. For such Hilbert spaces we have the following facts:

(e) The closure of $\mathfrak{C}_0(H_1,H_2)$ with respect to the operator norm (1) coincides with $\mathfrak{C}_\infty(H_1,H_2)$.

(f) $\mathfrak{C}_\infty(H)$ is a closed two-sided *-ideal in $\mathfrak{B}(H)$.

(g) The only closed two-sided ideals in $\mathfrak{B}(H)$ are: $\{0\}$, $\mathfrak{C}_\infty(H)$, $\mathfrak{B}(H)$.

1.2. The s-numbers of an operator. Given an operator $A \in \mathcal{B}(H_1,H_2)$ we define for $n \geqq 0$

$$s_n(A) = \inf\{\|A-F\| : F \in \mathfrak{C}_0(H_1,H_2),\ \dim F(H_1) \leqq n\}$$

and refer to the sequence $\{s_n(A)\}_{n=0}^{\infty}$ as the s-numbers of A. They have two remarkable properties:

(a) If $K \in \mathfrak{C}_\infty(H_1,H_2)$ then $\{s_n(K)\}_{n=0}^{\infty}$ is just the sequence of the eigenvalues (counted up to algebraic multiplicity) of $(K^*K)^{1/2}$. Note that the spectrum of a compact operator is at most countable, has at most one limit point, namely, 0, and each nonzero λ in the spectrum is an eigenvalue.

(b) (Horn's lemma) If $K,L \in \mathfrak{C}_\infty(H_1,H_2)$, $n \geqq 0$, and $1 \leqq p < \infty$, then

$$\sum_{j=0}^{n} s_j^p(KL) \leqq \sum_{j=0}^{n} s_j^p(K) \cdot s_j^p(L).$$

1.3. The Schatten - von Neumann classes. This are the classes $\mathfrak{C}_p(H_1,H_2)$, $1 \leqq p < \infty$, consisting by definition of all operators $K \in \mathcal{B}(H_1,H_2)$ for which

$$\|K\|_p := \Big(\sum_{n\geqq 0} s_n^p(K)\Big)^{1/p} < \infty. \tag{2}$$

Note that

$$\mathfrak{C}_\infty(H_1,H_2) = \{K \in \mathcal{B}(H_1,H_2) : \lim_{n\to\infty} s_n(K) = 0\}.$$

Moreover, $\|K\|_\infty := \sup_{n \geqq 0} s_n(K)$ $(= s_0(K))$ coincides with the operator norm of K defined by (1).

We list some properties of the classes $\mathfrak{C}_p(H_1,H_2)$:

(a) $\mathfrak{C}_p(H_1,H_2)$, $1 \leqq p < \infty$, is a Banach space under the norm (2). For $1 \leqq r < s \leqq \infty$ the space $\mathfrak{C}_r(H_1,H_2)$ is continuously and densely embedded in $\mathfrak{C}_s(H_1,H_2)$.
The set $\mathfrak{C}_0(H_1,H_2)$ is dense in $\mathfrak{C}_p(H_1,H_2)$ for every $1 \leqq p \leqq \infty$.

(b) $\mathfrak{C}_p(H)$, $1 \leqq p \leqq \infty$, is a two-sided (for $p < \infty$, in general, not closed) *-ideal in $\mathcal{B}(H)$. We have $\|K^*\|_p = \|K\|_p$.

(c) If $K \in \mathfrak{C}_p(H)$, $L \in \mathfrak{C}_q(H)$, $1 \leqq p,q,r \leqq \infty$, $1/r = 1/p + 1/q$, then $KL \in \mathfrak{C}_r(H)$ and

$$\|KL\|_r \leqq \|K\|_p\|L\|_q.$$

(d) Suppose $\{B_n\}$ and $\{C_n\}$ are two sequences of bounded operators such that $B_n \to B$ strongly and $C_n^* \to C^*$ strongly.
Then if $K \in \mathfrak{C}_p(H)$ $(1 \leqq p \leqq \infty)$

$$\|B_nKC_n - BKC\|_p \to 0 \quad (n \to \infty).$$

For every positive self-adjoint operator $K \in \mathfrak{B}(H)$ and any orthonormal basis $\{e_n\}_{n\geqq 0}$ in H the sum $\sum_{n\geqq 0}(Ke_n, e_n)$ can be shown to be independent of the choice of $\{e_n\}_{n\geqq 0}$. This sum is denoted by $\operatorname{tr}K$.

(e) $K \in \mathfrak{C}_p(H)$, $1 \leq p < \infty$, if and only if $\operatorname{tr}(|K|^p) < \infty$, where $|K| := (K^*K)^{1/2}$. In that case

$$\|K\|_p = \Big(\sum_{n\geqq 0} s_n^p(K)\Big)^{1/p} = \operatorname{tr}(|K|^p).$$

(f) For $K \in \mathfrak{C}_p(H)$, $1 \leqq p < \infty$, let $\{\lambda_n(K)\}_{n\geqq 0}$ denote the sequence of the eigenvalues (counted up to algebraic multiplicity). Then Weyl's inequality holds:

$$\sum_{n\geqq 0} |\lambda_n(K)|^p \leqq \|K\|_p^p.$$

1.4. <u>Trace class operators.</u> One can show that for every $K \in \mathfrak{C}_1(H)$ and for any orthonormal basis $\{e_n\}_{n\geqq 0}$ in H the series $\sum_{n\geqq 0}(Ke_n, e_n)$ converges absolutely and that its sum, denoted by $\operatorname{tr}K$, does not depend on the choice of $\{e_n\}_{n\geqq 0}$. The operators belonging to $\mathfrak{C}_1(H)$ are called <u>trace class operators</u>, and for $K \in \mathfrak{C}_1(H)$ the number $\operatorname{tr}K$ is referred to as the <u>trace</u> of K. The main properties of the trace are:

(a) $\operatorname{tr}(\cdot)$ is a *-linear functional on $\mathfrak{C}_1(H)$ and $|\operatorname{tr}K| \leqq \|K\|_1$ for every $K \in \mathfrak{C}_1(H)$.

(b) If either $A \in \mathfrak{C}_p(H)$, $B \in \mathfrak{C}_q(H)$, $1 < p < \infty$, $1/p + 1/q = 1$, or $A \in \mathfrak{C}_1(H)$, $B \in \mathfrak{B}(H)$, then $\operatorname{tr}(AB) = \operatorname{tr}(BA)$.

(c) If $K \in \mathfrak{C}_1(H)$ and if $A \in \mathfrak{B}(H)$ is invertible, then $\operatorname{tr}(AKA^{-1}) = \operatorname{tr}K$.

1.5. <u>Hilbert-Schmidt operators.</u> The operators of $\mathfrak{C}_2(H_1, H_2)$ are called <u>Hilbert-Schmidt operators.</u> They admit the following simple characterization:

Let $\{e_n\}_{n\geqq 0}$ and $\{f_n\}_{n\geqq 0}$ be arbitrary orthonormal bases in H_1 and H_2, respectively. Then

$$K \in \mathfrak{C}_2(H_1, H_2) \iff \sum_{n\geqq 0} \|Ke_n\|^2 < \infty$$

$$\iff \sum_{n\geqq 0} |(Ke_n, f_n)|^2 < \infty.$$

<u>1.6. Definition.</u> Suppose H is a finite-dimensional Hilbert space, $\dim H = n$, and let A be a linear operator on H. We write A in the form $A = I + K$. Denote by $\lambda_j(K)$ $(j = 1,\dots,n)$ the eigenvalues of K. Then $\{1 + \lambda_j(K)\}_{j=1}^{n}$ is the sequence of the eigenvalues of A and therefore,

$$\det A = \det(I + K) = \prod_{j=1}^{n} (1 + \lambda_j(K)).$$

Thus it seems reasonable to give the following definition.
Suppose now H is a separable Hilbert space and let $A \in \mathfrak{B}(H)$ be an operator of the form $A = I + K$ where $K \in \mathfrak{C}_1(H)$. If $\{\lambda_j(K)\}_{j \geq 0}$ denotes the sequence of the nonzero eigenvalues of K (counted up to algebraic multiplicity), then, by 1.3. (f), $\sum_{j \geq 0} |\lambda_j(K)| < \infty$. Hence the (possibly, infinite) product $\prod_{j \geq 0} (1 + \lambda_j(K))$ is absolutely convergent. We define

$$\det A = \det(I + K) := \prod_{j \geq 0} (1 + \lambda_j(K)).$$

In the case where the spectrum of K consists only of 0 we put $\det(I + K) = 1$. The number $\det A$ is called the <u>determinant</u> of A.

<u>1.7. Remark.</u> Note that besides the just given definition there are at least two other possibilities to define $\det(I + K)$ for abstract $K \in \mathfrak{C}_1(H)$, which, however, are in a certain sense equivalent to definition 1.6. (cf. Dunford, Schwartz [1] and Simon [1]). The technical details of any systematical treatment of operator determinants, i.e. the derivation of such of their properties which are analogous to the finite-dimensional situation, depend essentially on the definition of $\det(I + K)$ adopted in each case. Here we prefer the definition 1.6., due to Gohberg, Krein [2], for its simplicity, but we should note that the perhaps simplest and most elegant treatment of operator determinants is based on an other definition (see Simon [1]).

<u>1.8. Basic properties of operator determinants.</u> Here we list some properties of operator determinants which we shall need in chapter 6. For the proofs we refer to Gohberg, Krein [2] or Simon [1].

(a) If $K \in \mathfrak{C}_1(H)$ then

$$|\det(I + K)| \leq \exp(\|K\|_1),$$

$$|\det(I + K)| \leq \prod_{j \geq 0} (1 + s_j(K)).$$

The mapping defined by

$$\mathfrak{C}_1(H) \to \mathbb{C}, \quad K \mapsto \det(I + K),$$

is continuous, i.e. $\|K_n - K\|_1 \to 0$ $(n \to \infty)$ implies

$$|\det(I + K_n) - \det(I + K)| \to 0 \quad (n \to \infty).$$

Moreover, for $K,L \in \mathfrak{C}_1(H)$ we have

$$|\det(I + K) - \det(I + L)| \leqq \|K-L\|_1 \exp(\|K\|_1 + \|L\|_1 + 1).$$

(b) If $K(\lambda)$ is a $\mathfrak{C}_1(H)$-valued analytic function in $D \subset \mathbb{C}$, then $\det(I + K(\lambda))$ is analytic in D, too.

(c) If $K,L \in \mathfrak{C}_1(H)$ then

$$\det(I + K)\det(I + L) = \det(I + K + L + KL).$$

Suppose $K \in \mathfrak{C}_\infty(H)$, $A \in \mathfrak{B}(H)$, $AK \in \mathfrak{C}_1(H)$, $KA \in \mathfrak{C}_1(H)$. Then

$$\det(I + AK) = \det(I + KA).$$

In particular, for every $K \in \mathfrak{C}_1(H)$ and for every invertible operator $C \in \mathfrak{B}(H)$ we have

$$\det(I + CKC^{-1}) = \det(I + K).$$

(d) If P is a finite-rank projection then

$$\det P(I + K)P = \det(I + PKP),$$

where the det on the left refers to the ordinary finite-dimensional determinant for operators acting on Im P.

(e) If $K \in \mathfrak{C}_1(H)$ then $\det e^K = e^{\operatorname{tr} K}$ (note that $e^K - I = K + \frac{K^2}{2!} + \dots \in \mathfrak{C}_1(H)$).

(f) Let $K \in \mathfrak{C}_1(H)$. Then $\det(I + K) \neq 0$ if and only if $I + K$ is invertible in $\mathfrak{B}(H)$.

<u>1.9. Regularized determinants.</u> Suppose $K \in \mathfrak{C}_p(H)$, where $p > 1$ is an integer. An easy computation shows that in this case

$$R_p(K) := [(I+K)\exp(\sum_{j=1}^{p-1} (-K)^j/j)] - I \in \mathfrak{C}_1(H).$$

Thus it is justified to define

$$\det_p(I + K) := \det(I + R_p(K)).$$

We call $\det_p(I + K)$ the <u>p-regularized</u> <u>determinant</u> of $I + K$.

(a) Denoting by $\{\lambda_j(K)\}_{j\geqq 0}$ the sequence of the eigenvalues of $K \in \mathfrak{C}_p(H)$, we have

$$\det_p(I+K) = \prod_{j\geqq 0} [(1+\lambda_j(K))\exp(\sum_{l=1}^{p-1} (-\lambda_j(K))^l/l)].$$

(b) If $K \in \mathfrak{C}_p(H)$ then

$$\det_{p+1}(I+K) = \det_p(I+K) \cdot \exp[(-1)^p \operatorname{tr}(K^p)/p]$$

and, in particular, for $K \in \mathfrak{C}_1(H)$

$$\det_p(I+K) = \det(I+K) \cdot \exp[\sum_{j=1}^{p-1} (-1)^j \operatorname{tr}(K^j)/j].$$

(c) There exist constants Γ_p such that

$$|\det_p(I + K)| \leqq \exp(\Gamma_p \|K\|_p^p)$$

for every $K \in \mathfrak{C}_p(H)$. Thereby one can take $\Gamma_1 = 1$, $\Gamma_2 = 1/2$ and it holds $1/p \leqq \Gamma_p \leqq e(2 + \log p)$. The mapping given by

$$\mathfrak{C}_p(H) \to C, \quad K \mapsto \det_p(I + K),$$

is continuous. Moreover, for $K, L \in \mathfrak{C}_p(H)$,

$$|\det_p(I + K) - \det_p(I + L)|$$

$$\leqq \|K - L\|_p \exp[\Gamma_p(\|K\|_p + \|L\|_p + 1)^p].$$

(d) If $K, L \in \mathfrak{C}_2(H)$ then

$$\det_2(I+K)\det_2(I+L) = \det_2(I+K+L+KL) \exp(\operatorname{tr} KL).$$

(e) Suppose $K \in \mathfrak{C}_p(H)$. Then $\det_p(I + K) \neq 0$ if and only if $I + K$ is invertible in $\mathfrak{B}(H)$.

1.10. An identity for operator determinants. Later on we shall need the following remarkable fact, for whose proof we refer to Helton, Howe [1] and Pincus [1]:

Suppose $A, B \in \mathfrak{B}(H)$ and $AB - BA \in \mathfrak{C}_1(H)$. Then

$$\det(e^A e^B e^{-A} e^{-B}) = e^{\operatorname{tr}(AB - BA)}.$$

FREDHOLM OPERATORS

1.11. Definitions. Let X and Y be Banach spaces, and let $A \in \mathfrak{B}(X,Y)$. We put

$$\operatorname{Ker} A = \{x \in X : Ax = 0\}, \quad \operatorname{Im} A = A(X).$$

The operator A is said to be normally solvable if $\operatorname{Im} A$ is closed in Y. A normally solvable operator $A \in \mathfrak{B}(X,Y)$ is called Φ_+-operator (Φ_--operator), if $\dim \operatorname{Ker} A < \infty$ ($\dim \operatorname{Coker} A := \dim(Y/\operatorname{Im} A) < \infty$). If A is both a Φ_+- and a Φ_--operator, then it is called Φ-operator (or Fredholm operator or Noetherian operator) and the integer

$$\operatorname{Ind} A := \dim \operatorname{Ker} A - \dim \operatorname{Coker} A$$

is referred to as the index of A. The collection of all Φ_+-operators of X into Y is denoted by $\Phi_+(X,Y)$, and in the case $X = Y$ simply by $\Phi_+(X)$. Analogously the notations $\Phi(X,Y)$, $\Phi(X)$, $\Phi_-(X,Y)$, $\Phi_-(X)$ are to be understood.

1.12. Basic properties of Fredholm operators. The facts listed below are well-known.

(a) For $A \in \mathfrak{B}(X,Y)$ the following conditions are equivalent:

(i) $A \in \Phi(X,Y)$.

(ii) There exist operators $R,L \in \mathfrak{B}(X,Y)$ such that

$$AR - I_Y \in \mathfrak{C}_\infty(Y),\ LA - I_X \in \mathfrak{C}_\infty(X).$$

(iii) There exists an operator $B \in \mathfrak{B}(Y,X)$ such that

$$AB - I_Y \in \mathfrak{C}_0(Y),\ BA - I_X \in \mathfrak{C}_0(X).$$

Here I_X and I_Y are the identity operators in X and Y, respectively.

Any operator $B \in \mathfrak{B}(Y,X)$ for which

$$AB - I_Y \in \mathfrak{C}_\infty(Y),\ BA - I_X \in \mathfrak{C}_\infty(X)$$

is called a regularizer of A.

(b) Let $A \in \Phi(X,Y)$ and let B be an operator satisfying condition (iii) of (a). Then

$$\operatorname{Ind} A = \operatorname{tr}(I_X - BA) - \operatorname{tr}(I_Y - AB).$$

In particular, for $X = Y$

$$\operatorname{Ind} A = \operatorname{tr}(AB - BA).$$

(c) Suppose X, Y, Z are Banach spaces, and let $A \in \Phi(X,Y)$, $B \in \Phi(Y,Z)$. Then $BA \in \Phi(X,Z)$ and

$$\operatorname{Ind} BA = \operatorname{Ind} B + \operatorname{Ind} A.$$

(d) $\Phi(X,Y)$ is an open subset of $\mathfrak{B}(X,Y)$ and the mapping $\operatorname{Ind} : \Phi(X,Y) \to \mathbb{Z}$ is constant on the connected components of $\Phi(X,Y)$. From $A \in \Phi(X,Y)$, $K \in \mathfrak{C}_\infty(X,Y)$ it follows $A + K \in \Phi(X,Y)$ and $\operatorname{Ind}(A + K) = \operatorname{Ind} A$.

(e) Suppose X, Y, Z are Banach spaces, and let $A \in \mathfrak{B}(X,Y)$, $B \in \mathfrak{B}(Y,Z)$. Then the following implications hold:

(i) $A \in \Phi_\pm(X,Y)$, $B \in \Phi_\pm(Y,Z) \Rightarrow BA \in \Phi_\pm(X,Z)$.

(ii) $BA \in \Phi_+(X,Z) \Rightarrow A \in \Phi_+(X,Y)$.

(iii) $BA \in \Phi_-(X,Z) \Rightarrow B \in \Phi_-(Y,Z)$.

(iv) $A \in \Phi(X,Y)$, $BA \in \Phi(X,Z) \Rightarrow B \in \Phi(Y,Z)$.

(v) $B \in \Phi(Y,Z)$, $BA \in \Phi(X,Z) \Rightarrow A \in \Phi(X,Y)$.

(f) Property (d) remains valid, if one replaces Φ by Φ_+ or Φ_- and the index is allowed to assume the values $\pm\infty$. Moreover, for $A \in \Phi_\pm(X,Y)$

$$\dim \operatorname{Ker}(A+C) \leq \dim \operatorname{Ker} A, \quad \dim \operatorname{Coker}(A+C) \leq \dim \operatorname{Coker} A,$$

if only $C \in \mathfrak{B}(X,Y)$ has sufficiently small norm;

(g) Let $A \in \Phi_+(X)$ and let $K \in \mathfrak{C}_\infty(X)$ be the projection (with finite rank) onto $\operatorname{Ker} A$. Then there is a $\delta > 0$ such that

$$\|Ax\| + \|Kx\| \geq \delta\|x\|$$

for all $x \in X$. Vice versa, if for a given operator $A \in \mathfrak{B}(X)$ there exist operators $K_j \in \mathfrak{C}_\infty(X)$ $(j = 1,\dots,n)$ and a $\delta > 0$ such that

$$\|Ax\| + \sum_{j=1}^{n}\|K_jx\| \geq \delta\|x\|$$

for all $x \in X$, then $A \in \Phi_+(X)$.

A LOCAL PRINCIPLE

<u>1.13. Motivation.</u> Let X be a Banach space. By 1.1., $\mathfrak{C}_\infty(X)$ is a closed two-sided ideal in $\mathfrak{B}(X)$ and therefore $\mathfrak{B}(X)/\mathfrak{C}_\infty(X)$ is a Banach algebra (the so-called <u>Calkin algebra</u>). Denote by π the canonical projection of $\mathfrak{B}(X)$ onto $\mathfrak{B}(X)/\mathfrak{C}_\infty(X)$. From 1.12. (a) it follows that $A \in \Phi(X)$ if and only if πA is invertible in $\mathfrak{B}(X)/\mathfrak{C}_\infty(X)$. Thus the question of Fredholmness has been reduced to that of invertibility in a certain Banach algebra.

But there is a very effective tool to investigate invertibility in Banach algebras: the so-called local principles. We are now going to formulate the local principle of Gohberg and Krupnik, which is distinguished for its simplicity on the one hand and its wide area of applications on the other hand.

<u>1.14. The local principle of Gohberg and Krupnik.</u> Let $\mathcal{A}$ be a Banach algebra with identity e. A set $M \subset \mathcal{A}$ is called a <u>localizing class</u> if $0 \notin M$ and for any $f_1, f_2 \in M$ there exists a third element $f \in M$ such that $f_jf = ff_j = f$ $(j = 1,2)$. Two elements $a,b \in \mathcal{A}$ are said to be <u>M-equivalent</u>, written $a \overset{M}{\sim} b$, if

$$\inf_{f\in M}\|(a-b)f\| = \inf_{f\in M}\|f(a-b)\| = 0.$$

An element $a \in \mathcal{A}$ is called <u>M-invertible</u> if there exist $b \in \mathcal{A}$ and $f \in M$ such that $baf = fab = f$.

Finally, a system $\{M_\tau\}_{\tau\in T}$ of localizing classes is said to be <u>covering</u> if from each choice $\{f_\tau\}_{\tau\in T}$ $(f_\tau \in M_\tau)$ there can be selected a finite number of elements $f_{\tau_1},\dots,f_{\tau_m}$ whose sum is invertible in $\mathcal{A}$.

Theorem. Let $\{M_\tau\}_{\tau\in T}$ be a covering system of localizing classes, and let $b \in \mathcal{A}$ be M_τ-equivalent to $b_\tau \in \mathcal{A}$ $(\tau \in T)$. If b commutes with all elements of the set $\bigcup_{\tau\in T} M_\tau$, then b is invertible in $\mathcal{A}$ if and only if b_τ is M_τ-invertible for all $\tau \in T$.

Henceforth we denote by $G\mathcal{A}$ the group of invertible elements in a Banach algebra $\mathcal{A}$ with identity.

OPERATOR MATRICES AND THEIR DETERMINANTS

1.15. If X is any linear space, then X_n will denote the linear space of column-vectors of length n with entries from X, and $X_{n\times n}$ the linear space of $n\times n$-matrices with entries from X. For any Banach space X, the spaces X_n and $X_{n\times n}$ become Banach spaces if one introduces the norms by $\|(x_1,\dots,x_n)^T\|_{X_n} = \sum_{i=1}^{n} \|x_i\|_X$ and $\|(x_{jk})_{j,k=1}^{n}\|_{X_{n\times n}} = \max_k \sum_{j=1}^{n} \|x_{jk}\|_X$, respectively.

Note that the spaces $\mathcal{B}(X_n)$ and $(\mathcal{B}(X))_{n\times n}$ are isometrically isomorphic; in particular, every operator $\mathcal{A} \in \mathcal{B}(X_n)$ may be written as an operator matrix $\mathcal{A} = (A_{jk})_{j,k=1}^{n}$, $A_{jk} \in \mathcal{B}(X)$.

The question we are concerned with here is that of the connection between the Fredholm properties of operator matrices and their determinants. Given an operator matrix $\mathcal{A} \in (\mathcal{B}(X))_{n\times n}$ we form its determinant $\det \mathcal{A} \in \mathcal{B}(X)$ in the usual way, for instance,

$$\det(A_{jk})_{j,k=1}^{n} = \sum_{\sigma} (-1)^{p(\sigma)} A_{1,\sigma(1)} \cdots A_{n,\sigma(n)}, \tag{3}$$

where σ runs through all permutations of $\{1,\dots,n\}$ and where $p(\sigma)$ denotes the signum of the permutation σ. In the theorems following below, however, it will not be of significance, in which order the (in general, not commuting) factors in each item of the sum (3) are arranged.

Finally, the commutator $AB - BA$ will henceforth be denoted by $[A,B]$.

Theorem. Let X be a Banach space, and let

$$\mathcal{A} = (A_{jk})_{j,k=1}^{n} \in \mathcal{B}(X_n).$$

(a) If $[A_{ij}, A_{lm}] \in \mathfrak{C}_\infty(X)$ for $j \neq m$ (resp. $i \neq l$) and $\mathcal{A} \in \Phi_+(X_n)$ (resp. $\Phi_-(X_n)$), then $\det \mathcal{A} \in \Phi_+(X)$ (resp. $\Phi_-(X)$).

(b) If $[A_{ij}, A_{lm}] \in \mathfrak{C}_\infty(X)$ for $j \neq m$ (resp. $i \neq l$) and $\det \mathcal{A} \in \Phi_-(X)$ (resp. $\Phi_+(X)$), then $\mathcal{A} \in \Phi_-(X_n)$ (resp. $\Phi_+(X_n)$).

Corollary. If the entries of an operator matrix $A \in \mathcal{B}(X_n)$ pairwise commute modulo compact operators, then $A \in \Phi(X_n)$ (resp. $\Phi_+(X_n)$, $\Phi_-(X_n)$) if and only if $\det A \in \Phi(X)$ (resp. $\Phi_+(X)$, $\Phi_-(X)$).

1.16. Our next concern is the question of the relation between Ind A and Ind det A. There are examples showing that under the hypotheses of corollary 1.15. both equality and inequality of these indexes are possible. The following theorem gives a sufficient condition for equality.

Theorem. Let H be a Hilbert space, let $A = (A_{jk})_{j,k=1}^n \in \Phi(H_n)$ and suppose $[A_{jk}, A_{lm}] \in \mathcal{C}_1(H)$ for $j,k,l,m = 1,\dots,n$. Then $\det A \in \Phi(H)$ and

$$\operatorname{Ind} A = \operatorname{Ind} \det A.$$

Note that this theorem remains true in case H is a Banach space, but $[A_{jk}, A_{lm}] \in \mathcal{C}_0(H)$ for $j,k,l,m = 1,\dots,n$.

FUNCTIONS ON THE UNIT CIRCLE

1.17. The spaces L^p, H^p, C and C_A. Let Γ be the complex unit circle. We denote by L^p $(1 \leqq p < \infty)$ the Banach space of all (classes of) complex-valued measurable functions on Γ being summable (with respect to Lebesgue measure) in the p th power.
By C (resp. L^∞) we denote the Banach algebra of all continuous (resp. (classes of) measurable and essentially bounded) functions on Γ. Given $f \in L^1$ we define its Fourier coefficients f_n $(n \in \mathbb{Z})$ by

$$f_n = \frac{1}{2\pi} \int_0^{2\pi} f(e^{i\theta}) e^{-in\theta} d\theta.$$

Now, for $1 \leqq p \leqq \infty$, put

$$H^p = \{f \in L^p : f_n = 0 \text{ for all } n < 0\}$$

and

$$C_A = \{f \in C : f_n = 0 \text{ for all } n < 0\}.$$

Thus H^p (the so-called Hardy spaces) and C_A are closed subspaces of L^p and C, respectively. Moreover, H^∞ and C_A are Banach algebras.
The functions of L^2 and H^2 admit a very simple characterization via Fourier coefficients:

$$f \in L^2 (\text{resp. } H^2) \iff \sum_{n \in \mathbb{Z}} |f_n|^2 < \infty \; (\text{resp. } \sum_{n \in \mathbb{Z}_+} |f_n|^2 < \infty).$$

We define the projection $P : L^p \to H^p$ $(1 \leqq p \leqq \infty)$, the so-called Riesz projection, by

$$(Pf)_n = \begin{cases} f_n, & n \geqq 0 \\ 0, & n < 0. \end{cases}$$

Then P is bounded if and only if $1 < p < \infty$. Put $Q = I - P$ and $H^p_- = Q(L^p)$. Note that H^2_- is the orthogonal complement of H^2 in L^2, and that P and Q are orthogonal projections in the case $p = 2$.

Each function $\varphi \in H^1$ admits an analytic extension, denoted by $\hat{\varphi}$, into the interior $\mathbb{D}$ of the unit circle Γ:

$$\hat{\varphi}(z) = \sum_{n \in \mathbb{Z}_+} \varphi_n z^n \quad (z \in \mathbb{D}).$$

Given a function $\varphi \in L^1$ we denote by $\hat{\varphi}$ its harmonic extension into $\mathbb{D}$, that is,

$$\hat{\varphi}(re^{i\vartheta}) = \sum_{n \in \mathbb{Z}} \varphi_n r^{|n|} e^{in\vartheta} \quad (0 \leqq r < 1,\ 0 \leqq \vartheta < 2\pi).$$

Note that the harmonic extension has a representation via the Poisson integral,

$$\hat{\varphi}(re^{i\vartheta}) = \frac{1}{2\pi} \int_0^{2\pi} k_r(\vartheta - t)\varphi(e^{it})dt,$$

where $k_r(t) = (1 - r^2)/(1 - 2r \cos t + r^2)$.

An important property of functions from H^2 is reflected by the following theorem.

Theorem (F. and M. Riesz). A function in H^2 vanishes either almost everywhere or almost nowhere.

1.18. Inner and outer functions. A function $\varphi \in H^\infty$ is said to be an inner function if $|\varphi| = 1$ a.e. We denote by $\mathcal{P}_+$ the collection of all (trigonometric) polynomials from H^∞. A function $\varphi \in H^2$ is said to be an outer function if $E_\varphi := \text{clos}_{H^2}\{\varphi \cdot f : f \in \mathcal{P}_+\} = H^2$. Note that an outer function φ satisfies necessarily $\hat{\varphi}(z) \not\equiv 0 (z \in \mathbb{D})$, but that this condition is not sufficient for φ to be an outer function.

Theorem. Let $\varphi \in H^\infty$. Then the following statements are equivalent:

(a) $\varphi \in GH^\infty$.

(b) $\inf_{z \in \mathbb{D}} |\hat{\varphi}(z)| > 0$.

(c) $\varphi \in GL^\infty$ and φ is an outer function.

The following lemma will be applied in chapter 2.

Lemma. Suppose $f \in L^1$ is a positive function satisfying $\log f \in L^1$. Then there exists an outer function $h \in H^2$ such that $f = |h|^2$.

If $f \in L^2$, $|f| \geqq \varepsilon > 0$ a.e., then there is an outer function $g \in H^2$ such that $|f| = |g|$.

1.19. <u>The algebra $C + H^\infty$.</u> Define

$$C + H^\infty := \{f + h : f \in C,\ h \in H^\infty\}.$$

It turns out that this class of functions has a series of remarkable properties.

<u>Theorem.</u> $C + H^\infty$ is a closed subalgebra of L^∞. If A is any Banach algebra (with respect to the norm of L^∞) such that $H^\infty \subset A \subset L^\infty$, then either $A = H^\infty$ or $C + H^\infty \subset A$.

The Banach algebra $C + H^\infty$ is usually called the <u>Douglas algebra.</u>

<u>Theorem.</u> Let $\varphi \in C + H^\infty$. Then the following two statements are equivalent:

(a) $\varphi \in G(C + H^\infty)$.

(b) $\hat{\varphi}$ is bounded away from zero in some annulus near Γ, i.e. there exist $\delta > 0$ and $\varepsilon > 0$ such that

$$|\hat{\varphi}(re^{i\vartheta})| \geqq \varepsilon \quad \text{for} \quad 1 - \delta < r < 1, \quad 0 \leqq \vartheta < 2\pi.$$

Therefore, an invertible element in $C + H^\infty$ has a well-defined index

$$\operatorname{ind} \varphi = \frac{1}{2\pi}[\arg \hat{\varphi}(re^{i\vartheta})]_{\vartheta=0}^{2\pi},$$

which is independent of $r > 1 - \delta$. A further important property of $C + H^\infty$ is the so-called <u>asymptotic multiplicativity</u> of the harmonic extension on $C + H^\infty$.

<u>Theorem.</u> For $\varphi \in C + H^\infty$ and $0 < r < 1$, $0 \leqq \vartheta < 2\pi$ define $\varphi_r(e^{i\vartheta}) = \hat{\varphi}(re^{i\vartheta})$. Then if $f, g \in C + H^\infty$

$$\|f_r g_r - (fg)_r\|_{L^\infty} \to 0 \quad \text{as} \quad r \to 1.$$

1.20. <u>The Hölder classes C^α and the Besov classes B_p^α.</u> For $\alpha > 0$ and $1 \leqq p \leqq \infty$ we define the <u>Besov class</u> B_p^α by

$$B_p^\alpha = \{f \in L^p : \int_0^{2\pi} \frac{\|\Delta_t^n f\|_{L^p}^p}{|t|^{1+\alpha p}}\, dt < \infty\} \quad (p < \infty),$$

$$B_\infty^\alpha = \{f \in L^\infty : \sup_{t>0} \frac{\|\Delta_t^n f\|_{L^\infty}}{|t|^\alpha} < \infty\} \qquad (p = \infty),$$

where n is an integer such that $\alpha < n$, and we adopted the notation $\Delta_t^n = \Delta_t \Delta_t^{n-1}$, $(\Delta_t f)(e^{i\vartheta}) = f(e^{i(\vartheta+t)}) - f(e^{i\vartheta})$ $(\vartheta, t \in \mathbb{R})$. Note that this definition does not depend on the choice of n, $n > \alpha$. The classes B^{α}_{∞} coincide with the familar Hölder-Zygmund classes, i.e.,

$$B^{\alpha}_{\infty} = \{f \in C : |f(t_1)-f(t_2)| \leqq M_f|t_1-t_2|^{\alpha} \quad t_1,t_2 \in \Gamma\} \ (0 < \alpha < 1),$$

$$B^{1}_{\infty} = \{f \in C : |f(e^{i(\vartheta+t)}) + f(e^{i(\vartheta-t)}) - 2f(e^{i\vartheta})| \leqq$$

$$\leqq M_f|t| \quad \vartheta, t \in \mathbb{R}\},$$

and denoting by C^n, where n is a positive integer, the class of n-times continuously differentiable functions on Γ, we have finally

$$B^{\beta}_{\infty} = \{f \in C^n : f^{(n)} \in B^{\beta-n}_{\infty}\} \quad (n < \beta \leqq n + 1).$$

For this reason we shall henceforth write C^{α} instead of B^{α}_{∞}. The closure of the set of polynomials in C^{α} will be denoted by c^{α}.
We define the operator of fractional integration I_s $(s \in \mathbb{R})$ on the space of distributions on Γ by

$$(I_s f)_n = (1 + |n|)^{-s} f_n \quad (n \in \mathbb{Z}).$$

Then if only $\alpha > 0$, $\alpha + s > 0$

$$I_s(B^{\alpha}_p) = B^{\alpha+s}_p \,. \tag{4}$$

Clearly, (4) can be used to define B^{α}_p for $\alpha \leqq 0$. Provided with the usual norm B^{α}_p becomes a Banach space.
If $p = 2$ then

$$f \in B^{\alpha}_2 \iff \sum_{n \in \mathbb{Z}} (1 + |n|)^{2\alpha} |f_n|^2 < \infty.$$

Finally, put

$$(B^{\alpha}_p)_A = \{f \in B_p : f_n = 0 \text{ for } n < 0\}.$$

Thus $(B^{\alpha}_p)_A$ is a closed subspace of B^{α}_p. For $\alpha > 0$ this space can be characterized as the class of all functions $f \in L^p$ such that

$$\int_0^1 (1 - r)^{(n-\alpha)p-1} \|f_r^{(n)}\|^p_{L^p} dr < \infty \quad (p < \infty),$$

$$\sup_{0<r<1} (1 - r)^{n-\alpha} \|f_r^{(n)}\|_{L^\infty} < \infty \quad (p = \infty),$$

where n is an integer satisfying $\alpha < n$ and

$$f_r^{(n)}(e^{i\vartheta}) := \hat{f}^{(n)}(re^{i\vartheta}), \quad 0 \leqq \vartheta < 2\pi.$$

NOTES AND COMMENTS

1.1. - 1.9. Proofs of the facts presented here may be found in

Gohberg, Krein [2], Reed, Simon [1], Dunford, Schwartz [1], and in the excellent article Simon [1]. Proposition 1.3(d) is taken from Widom [8].

1.10. See Helton, Howe [1] and Pincus [1].

1.11. - 1.12. The theory of Fredholm operators is covered by many recent textbooks. We cite Douglas [2], Rudin [2], Gohberg, Krupnik [2], Prößdorf [1] and Michlin, Prößdorf [1].

1.13. - 1.14. Local principles were first fully utilized to investigate convolutions and related operators by Simonenko in [3], [4], [5]. Subsequently a host of local principles was established, see Gohberg, Krupnik [2], Kozak [1], [2], Pilidi [1], Douglas [2], [3], and Krupnik [1], although there is a close relationship between all of these principles. A discussion of this question for the case of one-dimensional singular integral operators may be found in Clancey, Gohberg [1], but we do not know any reference dealing with the comparization of several local theories from a general point of view.
We prefer the local principle of Gohberg and Krupnik for its simplicity in applications.

1.15. These results are essentially due to Köhler and Silbermann [1], [2], but were independently also found by Markus and Feldman [1].

1.16. This theorem goes back to Markus and Feldman [2].

1.17. - 1.19. We refer to Rudin [1], Douglas [2], Hoffman [1], Duren [1], Sarason [1] as standard works.

1.20. These facts are taken from Peller [3] and Peller, Hruščev [1]. For proofs we recommend the reader to consult Stein [1], Triebel [1], Bergh, Löfström [1].

CHAPTER 2

TOEPLITZ OPERATORS

DEFINITION OF TOEPLITZ AND HANKEL OPERATORS

2.1. Multiplication and Toeplitz operators. Let $\tilde{l}^p$, $1 \leqq p < \infty$, be the Banach space of all sequences $\varphi = \{\varphi_j\}_{j\in\mathbb{Z}}$ of complex numbers satisfying

$$\|\varphi\|_p = \Big(\sum_{j\in\mathbb{Z}} |\varphi_j|^p\Big)^{1/p} < \infty.$$

By l^p we denote the closed subspace of $\tilde{l}^p$ defined by

$$l^p = \{\varphi \in \tilde{l}^p : \varphi_j = 0 \quad \text{for} \quad j < 0\}.$$

Finally, let $\tilde{l}^o$ (resp. l^o) denote the collection of all sequences from $\tilde{l}^p$ (resp. l^p) that have a finite support.

Given $a \in L^1$ with Fourier coefficients $\{a_n\}_{n\in\mathbb{Z}}$ we define $M(a)$ and $T(a)$ on $\tilde{l}^o$ and l^o, respectively, as

$$M(a) : \{\varphi_j\}_{j\in\mathbb{Z}} \mapsto \{\sum_{j\in\mathbb{Z}} a_{i-j}\varphi_j\}_{i\in\mathbb{Z}}, \tag{1}$$

$$T(a) : \{\varphi_j\}_{j\in\mathbb{Z}_+} \mapsto \{\sum_{j\in\mathbb{Z}_+} a_{i-j}\varphi_j\}_{i\in\mathbb{Z}_+} . \tag{2}$$

If

$$\sup\{\|M(a)\varphi\|_p/\|\varphi\|_p : \varphi \in \tilde{l}^o, \ \varphi \neq 0\} < \infty$$

(resp.

$$\sup\{\|T(a)\varphi\|_p/\|\varphi\|_p : \varphi \in l^o, \ \varphi \neq 0\} < \infty),$$

then $M(a)$ (resp. $T(a)$) can be extended to a bounded operator on $\tilde{l}^p$ (resp. l^p). In this case we call $M(a) \in \mathcal{B}(\tilde{l}^p)$ the multiplication operator with symbol $a(t)$, and $T(a) \in \mathcal{B}(l^p)$ is referred to as the Toeplitz operator with symbol $a(t)$. In order to avoid confusion based on the use of the single letter "a", henceforth a function "a" given on the unit circle Γ will frequently be written as "a(t)".

Note that for $p = 2$ the operators $M(a)$ and $T(a)$ via the isometrical isomorphisms

$$\tilde{l}^2 \to L^2, \ \{\varphi_j\}_{j\in\mathbb{Z}} \mapsto \sum_{j\in\mathbb{Z}} \varphi_j t^j (|t| = 1), \tag{3}$$

$$l^2 \to H^2, \ \{\varphi_j\}_{j\in\mathbb{Z}_+} \mapsto \sum_{j\in\mathbb{Z}_+} \varphi_j t^j (|t| = 1), \tag{4}$$

are equivalent to the operators given by

$$L^2 \to L^2, \quad \varphi \mapsto a\varphi,$$

$$H^2 \to H^2, \quad \varphi \mapsto P(a\varphi),$$

(cf. 1.17). In what follows we shall frequently identify these operators, i.e. we think of $M(a)$ and $T(a)$ not only as acting on spaces of sequences, but also on the corresponding spaces of functions. Thus we identify $T(a)$ with the operator defined on H^2 by $PM(a)P$.

2.2. The classes M_p. We define M_p to consist of all functions $a \in L^1$ for which $M(a) \in \mathcal{B}(\Upsilon^p)$. Obviously, M_p is an algebra, and equipped with the norm $\|a\|_{M_p} := \|M(a)\|_{\mathcal{B}(\Upsilon_p)}$ it becomes a Banach algebra. The following facts are well-known:

$$M_1 = W := \{a \in L^1 : \sum_{n \in \mathbb{Z}} |a_n| < \infty\}$$

(W is usually called the Wiener algebra),

$$M_2 = L^\infty,$$

$$M_1 \subset M_p = M_q \subset M_2 \quad \text{for} \quad 1 < p < \infty, \quad 1/p + 1/q = 1,$$

and M_p $(1 < p < \infty)$ contains all functions $a(t)$ with finite total variation $V_1(a)$ on Γ. Thereby,

$$\|a\|_{M_p} \leqq c_p(\|a\|_{L^\infty} + V_1(a)) \tag{5}$$

with some constant c_p independent of $a(t)$.

2.3. Hankel operators. For $a \in M_p$ the Hankel operator $H(a) \in \mathcal{B}(l^p)$ is defined as

$$H(a) : \{\varphi_j\}_{j \in \mathbb{Z}_+} \mapsto \{\sum_{j \in \mathbb{Z}_+} a_{i+j+1}\varphi_j\}_{i \in \mathbb{Z}_+}. \tag{6}$$

Via 2.1(3), (4) in the case $p = 2$ the Hankel operator $H(a)$ can be thought of as acting on H^2 by the rule

$$\varphi \mapsto PM(a)QJ\varphi,$$

where P and Q are the projections introduced in 1.17, and where J is the reflexion operator defined by

$$(Jf)(t) = \frac{1}{t} f(\frac{1}{t}) = \sum_{j \in \mathbb{Z}} f_j \frac{1}{t^{j+1}} \quad (f \in L^2).$$

For $a \in M_p$ we put $\tilde{a}(t) = a(1/t)$ $(|t| = 1)$. This notation will be used throughout the following. Obviously, $a \in M_p$ implies $\tilde{a} \in M_p$, and the operators $T(\tilde{a})$ and $H(\tilde{a})$ act on l^p by the rules

$$T(\tilde{a}) : \{\varphi_j\}_{j \in \mathbb{Z}_+} \mapsto \{\sum_{j \in \mathbb{Z}_+} a_{j-i}\varphi_j\}_{i \in \mathbb{Z}_+},$$

$$H(\tilde{a}) : \{\varphi_j\}_{j \in \mathbb{Z}_+} \mapsto \{\sum_{j \in \mathbb{Z}_+} a_{-i-j-1}\varphi_j\}_{i \in \mathbb{Z}_+}.$$

Again via 2.1(3), (4) the operators $T(\tilde{a})$ and $H(\tilde{a})$ can be identified respectively with $JQM(a)QJ$ and $JQM(a)P$, both thought of as acting on H^2.

2.4. We are going to introduce still some notations. Let e_n, $n \in \mathbb{Z}$, be the element of $\tilde{l}^p$ which is given by $\{\delta_{jn}\}_{j\in\mathbb{Z}}$ (δ – the Kronecker symbol). The function χ_n, $n \in \mathbb{Z}$, is defined by $\chi_n(t) = t^n$ ($|t|=1$). Obviously, $\{e_n\}_{n\in\mathbb{Z}}$ (resp. $\{e_n\}_{n\in\mathbb{Z}_+}$ forms a basis in $\tilde{l}^p$ (resp. l^p) if only $1 \leqq p < \infty$. Furthermore, $\{\chi_n\}_{n\in\mathbb{Z}}$ (resp. $\{\chi_n\}_{n\in\mathbb{Z}_+}$ is an orthonormal basis in L^2 (resp. H^2).

Henceforth P will denote not only the Riesz projection of L^p onto H^p (see 1.17), but also the natural projection of $\tilde{l}^p$ onto l^p. In both situations, Q is defined as $I - P$.

L^∞ – SYMBOLS

2.5. Theorem (Brown- Halmos). If $T(a) \in \mathcal{B}(l^p)$, $1 < p < \infty$, then $M(a) \in \mathcal{B}(\tilde{l}^p)$, i.e. $a \in M_p$. Moreover,

$$\|T(a)\|_{\mathcal{B}(l^p)} = \|M(a)\|_{\mathcal{B}(\tilde{l}^p)}$$

and, in particular, for $p = 2$,

$$\|T(a)\|_{\mathcal{B}(l^2)} = \|a\|_{L^\infty} .$$

Proof. Put $U = M(\chi_1)$ (see 2.4). Then U is an isometry on $\tilde{l}^p$; therefore

$$\|T(a)\|_{\mathcal{B}(l^p)} = \|PM(a)P\|_{\mathcal{B}(\tilde{l}^p)} = \|U^{-n}PM(a)PU^n\|_{\mathcal{B}(\tilde{l}^p)}$$

for every $n > 0$. A simple computation shows that $U^{-n}PM(a)PU^n \to M(a)$ strongly in $\tilde{l}^p$. Thus by the Banach-Steinhaus theorem,

$$\|M(a)\|_{\mathcal{B}(\tilde{l}^p)} \leqq \|T(a)\|_{\mathcal{B}(l^p)}.$$

Now the other assertions of the theorem follow immediately. Δ

2.6. Proposition. Suppose $1 < p < \infty$, $1/p + 1/q = 1$ and let $a \in M_p$. We define $\bar{a}$ by $\bar{a}(t) = \overline{a(t)}$ ($|t| = 1$). Then

(a) $\bar{a} \in M_q$, and the adjoint operator $T^*(a) \in \mathcal{B}(l^q)$ of $T(a) \in \mathcal{B}(l^p)$ coincides with $T(\bar{a})$.

(b) $a \in M_r$ for every $r \in [p,q]$.

(c) $$\|a\|_{L^\infty} = \|T(a)\|_{\mathcal{B}(l^2)} \leqq \|T(a)\|_{\mathcal{B}(l^r)} = $$
$$= \|T(\bar{a})\|_{\mathcal{B}(l^r)} \leqq \|T(\bar{a})\|_{\mathcal{B}(l^p)} = \|T(a)\|_{\mathcal{B}(l^q)}$$
for every $r \in [p,q]$.

(d) $\|T(a)\|_{\mathcal{B}(l^r)} \leqq \|a\|_{L^\infty}^{1-\gamma} \|T(a)\|_{\mathcal{B}(l^p)}^{\gamma}$ for $r \in [p,q]$ and $\gamma = \dfrac{p|r-2|}{r|p-2|}$.

The corresponding properties for multiplication operators are well-known (cf. 2.2) or follow by standard arguments and by applying the Riesz-Thorin interpolation theorem. For the Toeplitz case it remains to apply Theorem 2.5. Δ

2.7. Proposition. If $a, b \in M_p$ $(1 \leq p < \infty)$ then

$$T(ab) = T(a)T(b) + H(a)H(\tilde{b}).$$

In particular, for $\overline{c_-} \in M_p \cap H^\infty$ and $d_+ \in M_p \cap H^\infty$,

$$T(c_-)T(a)T(d_+) = T(c_- a d_+).$$

Proof. We have

$$\begin{aligned} T(ab) &= PM(ab)P = PM(a)M(b)P \\ &= PM(a)PM(b)P + PM(a)QM(b)P \\ &= PM(a)\, P \cdot PM(b)P + PM(a)QJ \cdot JQM(b)P \end{aligned}$$

and it remains to recall 2.1 and 2.3. Δ

2.8. Theorem (Coburn). Let $a \in M_p$ $(1 < p < \infty)$ and suppose $a(t)$ is not identically zero. Then the kernel of $T(a)$ in l^p, or the kernel of $T(\overline{a})$ in l^q $(1/p + 1/q = 1)$ is trivial.

Proof. Let $f \in l^p$, $g \in l^q$, $T(a)f = 0$, $T(\overline{a})g = 0$. Define $f_+ \in \tilde{l}^p$ and $h_- \in \tilde{l}^q$ by

$$(f_+)_n := \begin{cases} f_n, & n \geq 0 \\ 0, & n < 0 \end{cases}, \qquad (h_-)_n := \begin{cases} 0, & n > 0 \\ \overline{g}_{-n}, & n \leq 0 \end{cases}.$$

Then $M(a)f_+ =: f_- \in \tilde{l}^p$ and $M(a)h_- =: h_+ \in \tilde{l}^q$, where $(f_-)_n = 0$ for $n \geq 0$ and $(h_+)_n = 0$ for $n \leq 0$. For $\varphi \in \tilde{l}^p$ and $\psi \in \tilde{l}^q$ we have $\varphi * \psi \in \tilde{l}^1$, where $(\varphi * \psi)_i = \sum_{j \in \mathbb{Z}} \varphi_{i-j}\psi_j$. Thus

$$\begin{aligned} f_- * h_- &= M(a)f_+ * h_- = a * f_+ * h_- \\ &= f_+ * a * h_- = f_+ * M(a)h_- = f_+ * h_+. \end{aligned}$$

Consequently, $f_- * h_- = f_+ * h_+ = 0$ and this gives $h_- = 0$, $f_+ = 0$ or $f_- = h_+ = 0$. In the first two cases we are done. Thus suppose $f_- = h_+ = 0$. Assume for a moment that $1 < p \leq 2$. From $f_+ \in l^p \subset l^2$ and $M(a)f_+ = 0$ we get $af_+ = 0$ $(a \in L^\infty, f_+ \in H^2)$. Since $a \not\equiv 0$, we can conclude that f_+ vanishes on a set of positive measure and the theorem from 1.17 leads to $f_+ = 0$. In the case $p \geq 2$ the same argument applied to the equality $M(a)h_- = 0$ gives $h_- = 0$, which completes the proof. Δ

Corollary. Let $a \in M_p$ $(1 < p < \infty)$, $T(a) \in \Phi(l^p)$ and $\operatorname{Ind} T(a) = 0$. Then $T(a) \in G\mathcal{B}(l^p)$.

Remark. The corollary remains true in the case $p = 1$, too. This follows from $M_1 = W$ and an application of the Wiener-Hopf factorization (see 5.24).

2.9. Definitions. If $\mathcal{A}$ is a Banach algebra with identity e, then the spectrum of an element $a \in \mathcal{A}$ is defined by

$$\sigma_{\mathcal{A}}(a) = \{\lambda \in \mathbb{C} : a - \lambda e \notin G\mathcal{A}\}.$$

If X is a Banach space and $A \in \mathcal{B}(X)$, then the spectrum $\sigma(A)$ of A is defined to be $\sigma_{\mathcal{B}(X)}(A)$, i.e.

$$\sigma(A) = \{\lambda \in \mathbb{C} : A - \lambda I \notin G\mathcal{B}(X)\},$$

and we refer to $\sigma_{\mathcal{B}(X)/\mathfrak{C}_\infty(X)}(\pi A)$ (cf. 1.13) as the essential spectrum $\sigma_{ess}(A)$ of A.
Equivalently,

$$\sigma_{ess}(A) = \{\lambda \in \mathbb{C} : A - \lambda I \notin \Phi(X)\}.$$

Note that $\sigma_{ess}(A) \subset \sigma(A)$.

Next, for $\varphi \in L^\infty$ we define the essential range $R(\varphi)$ of φ to consist of all $\lambda \in \mathbb{C}$ such that $\{t \in \Gamma : |\varphi(t) - \lambda| < \varepsilon\}$ has positive Lebesgue measure for every $\varepsilon > 0$. A little thought shows that

$$R(\varphi) = \sigma_{L^\infty}(\varphi) = \sigma_{\mathcal{B}(L^2)}(M(\varphi)).$$

Finally, the essential norm of an operator $A \in \mathcal{B}(X)$ is defined in the following way:

$$\|A\|_{ess} := \|\pi A\|_{\mathcal{B}(X)/\mathfrak{C}_\infty(X)} = \inf_{K \in \mathfrak{C}_\infty(X)} \|A + K\|_{\mathcal{B}(X)}.$$

Obviously, $\|A\|_{ess} \leqq \|A\|$.

2.10. Theorem (Hartman-Wintner). Let $a \in M_p$, $1 < p < \infty$. If $T(a) \in \Phi(l^p)$ then $a \in GM_p$ and, consequently, $a \in GL^\infty$. Thus

$$R(a) \subset \sigma_{ess}(T(a)).$$

Proof. Suppose $T(a) \in \Phi(l^p)$ and denote by K the projection (of finite rank) onto Ker A. By 1.12(g) there is a $\delta_1 > 0$ such that

$$\|T(a)\varphi\|_p + \|K\varphi\|_p \geqq \delta_1 \|\varphi\|_p$$

for all $\varphi \in l^p$. This implies

$$\|PM(a)P\psi\|_p + \|PKP\psi\|_p + \|Q\psi\|_p \geqq \delta_2 \|\psi\|_p$$

for all $\psi \in \tilde{l}^p$, and where $\delta_2 = \min\{\delta_1, 1\}$. Setting $U = M(\chi_1)$ (cf. 2.4 and the proof of Theorem 2.5) we get

$$\|PM(a)PU^n\varphi\|_p + \|PKPU^n\varphi\|_p + \|QU^n\varphi\|_p \geqq \delta_2 \|U^n\varphi\|_p$$

for all $\varphi \in \ell^p$. But U is an isometry, hence

$$\|U^{-n}PM(a)PU^n\varphi\|_p + \|U^{-n}PKPU^n\varphi\|_p + \|U^{-n}QU^n\varphi\|_p \geqq$$

$$\geqq \delta_2 \|U^{-n}U\varphi\| = \delta_2 \|\varphi\|_p$$

for all $\varphi \in \ell^p$. Since

$U^{-n}PM(a)PU^n \to M(a)$ strongly,

$U^{-n}PKPU^n \to 0$ strongly (note that $PU^n \to 0$ weakly),

$U^{-n}QU^n \to 0$ strongly,

we obtain that $\|M(a)\varphi\|_p \geqq \delta_2 \|\varphi\|_p$ for all $\varphi \in \ell^p$. Thus $M(a)$ has a trivial kernel and a closed image.
Analogously can be shown that $\|M(\overline{a})\varphi\|_q \geqq \delta_3 \|\varphi\|_q$ (with some $\delta_3 > 0$) for all $\varphi \in \ell^q$ $(1/p + 1/q = 1)$. Hence, the kernel of $M(\overline{a}) = M^*(a)$ in ℓ^q is trivial, too. Putting these things together we get the invertibility of $M(a)$ on l^p, which is equivalent to that of $a(t)$ in M_p. Δ

<u>Remark.</u> The theorem remains true for $p = 1$ (see the remark in 2.8).

<u>2.11.</u> Now we are going to deal mainly with the case $p = 2$. Note that in this case $T(a)$, $a \in L^\infty$, may be thought of as acting not only on l^2, but (even more advantageously) also on H^2.
Let us denote by $\mathrm{conv}(E)$ the convex hull of a set $E \subset \mathbb{C}$.

<u>Theorem (Brown-Halmos).</u> Let $a \in L^\infty$. Then

$$\sigma(T(a)) \subset \mathrm{conv}(R(a)). \tag{7}$$

<u>Proof.</u> Put $\mathbb{C}_+ := \{\zeta \in \mathbb{C} : \mathrm{Re}\zeta > 0\}$. To prove the theorem it will be sufficient to show the following implication:

$$R(a) \subset \mathbb{C}_+ \implies \sigma(T(a)) \subset \mathbb{C}_+.$$

But if $R(a) \subset \mathbb{C}_+$, then there is an $\varepsilon > 0$ such that $\varepsilon R(a) = R(\varepsilon a) \subset \{\zeta \in \mathbb{C} : |\zeta - 1| < r\}$ for some $r < 1$. Thus, by Theorem 2.5,

$$1 > \|\varepsilon a - \chi_0\|_{L^\infty} = \|\varepsilon T(a) - I\|_{\mathcal{B}(l^2)}.$$

This gives the invertibility of $T(a)$. Δ

<u>Remark.</u> In general, the inclusion (7) fails to be true for $p \neq 2$. One may easily find counterexamples among piecewise continuous symbols (see 2.31).

2.12. Combining Theorems 2.10 and 2.11 we get (for $p = 2$)

$$R(a) \subset \sigma_{ess}(T(a)) \subset \sigma(T(a)) \subset \mathrm{conv}(R(a)).$$

There are two special cases where the spectrum and the essential spectrum can be described completely. Since we shall not rely on these theorems in the sequel, we renounce to give the proofs and refer only to Douglas [2] and Nikol'skiĭ [3].

Theorem (Wintner). Let $a \in H^\infty$. Then

$$\sigma(T(a)) = \mathrm{clos}\, \hat{a}(D) = \{\lambda \in \mathbb{C} : a - \lambda \notin GH^\infty\},$$

$$\sigma_{ess}(T(a)) = R(a) \cup \{\lambda \in \mathbb{C} : \mathrm{codim}\, E_{a-\lambda} = \infty\}.$$

Here $\hat{a}$ is the analytic extension of $a \in H^\infty$ and for the definition of E_f see 1.18.

Theorem (Hartman-Wintner). Let $a(t)$ be a real-valued function in L^∞. Then

$$\sigma(T(a)) = \sigma_{ess}(T(a)) = [\mathrm{ess\,inf}\, a, \mathrm{ess\,sup}\, a].$$

Our next concern is the determination of the norm of a Hankel operator on H^2.

2.13. Theorem (Nehari). For $a \in L^\infty$

$$\|H(a)\|_{\mathcal{B}(H^2)} = \mathrm{dist}_{L^\infty}(\bar{a}, H^\infty).$$

Proof. Let p and q be trigonometric polynomials in H^2. An easy computation shows that $(H(a)p,q) = (H(a)\chi_0, p^*q)$, where $p^*(t) := \overline{p(1/t)}$. It is well-known that

$$\{f^* \cdot g : \|f\|_{H^2} \leqq 1, \|g\|_{H^2} \leqq 1\} = \{h \in H^1 : \|h\|_{H^1} \leqq 1\}.$$

Therefore,

$$\|H(a)\|_{\mathcal{B}(H^2)} = \sup\{(H(a)f,g) : \|f\|_{H^2} \leqq 1, \|g\|_{H^2} \leqq 1\}$$

$$= \sup\{\frac{1}{2\pi} \int_\Gamma H(a)\chi_0 \cdot h|dt| : \|\bar{h}\|_{H^1} \leqq 1\}$$

($|dt|$ denotes the Lebesgue measure on Γ) and, consequently, by

$$A : \overline{H^1} \to \mathbb{C}, \quad h \mapsto \frac{1}{2\pi} \int_\Gamma H(a)\chi_0 \cdot h|dt|$$

a linear functional with norm $\|A\| = \|H(a)\|$ is given. Consider all bounded extensions

$$\tilde{A} : L^1 \to \mathbb{C}, \quad f \mapsto \frac{1}{2\pi} \int_\Gamma \varphi \cdot f|dt|$$

of A.

We have $\|A\| \leq \|\chi\| = \|\varphi\|_{L^\infty}$. By the Hahn-Banach theorem there exists an extension χ_0 with $\|A\| = \|\chi_0\|$, and it results that

$$\|H(a)\| = \inf\|\chi\| = \inf\|\varphi\|_{L^\infty}.$$

But

$$\chi \text{ is an extension of } A$$

$$\iff \int_\Gamma (\varphi - H(a)\chi_0)\cdot h|dt| = 0 \qquad \forall\, h \in \overline{H^1}$$

$$\iff (\varphi - H(a)\chi_0)_n = 0 \qquad \forall\, n \geqq 0$$

$$\iff \chi_1\varphi - a \in \overline{H^\infty}.$$

Hence

$$\begin{aligned}\|H(a)\|_{\mathcal{B}(H^2)} &= \inf\{\|\varphi\|_{L^\infty} : \chi_1\varphi - a \in \overline{H^\infty}\}\\ &= \inf\{\|\chi_1\varphi\|_{L^\infty} : \chi_1\varphi - a \in \overline{H^\infty}\}\\ &= \operatorname{dist}_{L^\infty}(a, \overline{H^\infty}) = \operatorname{dist}_{L^\infty}(\bar{a}, H^\infty),\end{aligned}$$

and the proof is complete. Δ

The following proposition shows that the invertibility problem for Toeplitz operators can be reduced to that for the case of unimodular symbols.

<u>2.14. Proposition.</u> Let $a \in L^\infty$. Then

$$T(a) \in G\mathcal{B}(H^2) \iff a \in GL^\infty \text{ and } T(a/|a|) \in G\mathcal{B}(H^2).$$

<u>Proof.</u> If $a \in GL^\infty$ then, by the lemma in 1.18, there is an outer function $h \in H^2$ such that $|a^{-1}| = |h|^2$ and, consequently, $h \in H^\infty$. From Proposition 2.7 we obtain

$$T(a/|a|) = T(\bar{h}ah) = T^*(h)T(a)T(h). \tag{8}$$

Moreover, the theorem in 1.18 and Proposition 2.7 give that $T(h) \in G\mathcal{B}(H^2)$ and therefore also $T^*(h) \in G\mathcal{B}(H^2)$.
Now suppose $a \in GL^\infty$ and $T(a/|a|) \in G\mathcal{B}(H^2)$. Then (8) holds and the invertibility of $T(a)$ follows immediately.
Vice versa, assume $T(a) \in G\mathcal{B}(H^2)$. Then $a \in GL^\infty$ by Theorem 2.10. Thus (8) is again fulfilled and we get $T(a/|a|) \in G\mathcal{B}(H^2)$. Δ

<u>2.15. Theorem.</u> Let $u \in L^\infty$ be a unimodular function. Then

(a) $T(u)$ is left-invertible $\iff \operatorname{dist}_{L^\infty}(u, H^\infty) < 1$.

(b) $T(u)$ is right-invertible $\iff \operatorname{dist}_{L^\infty}(\bar{u}, H^\infty) < 1$.

(c) $T(u)$ is invertible $\iff$ there is an outer function $h \in H^\infty$ such that $\|u - h\|_\infty < 1$.

Proof.

(a) In view of $M(u)P = PM(u)P + QM(u)P$ and $|u| = 1$ we have

$$\|f\|^2 = \|uf\|^2 = \|P(uf)\|^2 + \|Q(uf)\|^2$$
$$= \|T(u)f\|^2 + \|H(\tilde{u})f\|^2$$

for every $f \in H^2$. Applying 1.12(g) we find that $T(u)$ is left-invertible if and only if there exists an $\varepsilon > 0$ such that $\varepsilon\|f\|^2 \leq \|T(u)f\|^2$ for every $f \in H^2$ holds. Consequently, $T(u)$ is left-invertible if and only if $\|H(\tilde{u})\| < 1$, which, by Theorem 2.13, is equivalent to

$$\mathrm{dist}_{L^\infty}(\bar{\tilde{u}}, H) = \mathrm{dist}_{L^\infty}(u, H^\infty) < 1.$$

(b) $T^*(u) = T(\bar{u})$.

(c) Suppose $T(u)$ is invertible. According to (a) there is an $h \in H^\infty$ such that $\|u - h\|_{L^\infty} < 1$. It remains to show that h is outer. From

$$\|I - T(\bar{u}h)\| = \|\chi_0 - \bar{u}h\|_{L^\infty} = \|u - h\|_{L^\infty} < 1 \qquad (9)$$

it follows the invertibility of

$$T(\bar{u}h) = T(\bar{u})T(h) = T^*(u)T(h). \qquad (10)$$

Thus $T(h)$ is also invertible and, in particular, surjective. Then 1.18 immediately implies that h is outer.
Now suppose $h \in H^\infty$ is an outer function such that $\|u - h\|_{L^\infty} < 1$. From (9) we get again the invertibility of (10). Since $|h(t)| \geq 1 - \|u - h\|_{L^\infty} > 0$, we obtain that $h \in GL^\infty$. Combining the theorem from 1.18 and Proposition 2.7 we can conclude that $T(h) \in G\mathcal{B}(H^2)$. Hence, by (2), $T(\bar{u}) \in G\mathcal{B}(H^2)$ and therefore $T(u) = T^*(\bar{u}) \in G\mathcal{B}(H^2)$. Δ

$C + H^\infty$ - SYMBOLS

2.16. Theorem. Let $a \in L^\infty$. Then

$$\|H(a)\|_{ess} = \mathrm{dist}_{L^\infty}(\bar{a}, C + H^\infty).$$

Consequently, $H(a) \in \mathfrak{C}_\infty(H^2)$ if and only if $\bar{a} \in C + H^\infty$ (or, equivalently, $a \in C + \overline{H^\infty}$).

The second assertion is usually referred to as Hartman's theorem.

Proof. If p is a trigonometric polynomial, then $H(p)$ is of finite rank. Thus from 1.1(e) we can conclude that $H(f) \in \mathfrak{C}_\infty(H^2)$ for every $f \in C$.

Therefore,

$$\operatorname{dist}_{L^\infty}(\overline{a}, C + H^\infty) = \inf_{f\in C, h\in H^\infty} \|\overline{a} - f - h\|_{L^\infty}$$

$$= \inf_{f\in C} \inf_{h\in H^\infty} \|\overline{a} - f - h\|_{L^\infty} = \inf_{f\in C} \operatorname{dist}_{L^\infty}(\overline{a} - f, H^\infty)$$

$$= \inf_{f\in C} \|H(a - \overline{f})\| \quad \text{(Theorem 2.13)}$$

$$= \inf_{f\in C} \|H(a) - H(\overline{f})\| \geqq \|H(a)\|_{ess}.$$

Now, put $V = T(\chi_1)$ (cf. 2.4). Since $(V^n)^* = T(\chi_{-n})$ converges strongly to zero as $n \to \infty$, we have $\|KV^n\| \to 0$ as $n \to \infty$ for every $K \in \mathfrak{C}_\infty(H^2)$ (see 1.3(d)). Thus

$$\|H(a) - K\| \geqq \|(H(a) - K)V^n\| \geqq \|H(a)V^n\| - \|KV^n\|$$

$$= \|H(\chi_{-n}a)\| - \|KV^n\|$$

$$= \operatorname{dist}_{L^\infty}(\chi_n\overline{a}, H^\infty) - \|KV^n\| \quad \text{(Theorem 2.13)}$$

$$= \operatorname{dist}_{L^\infty}(\overline{a}, \chi_{-n}H^\infty) - \|KV^n\|$$

$$\geqq \operatorname{dist}_{L^\infty}(\overline{a}, C + H^\infty) - \|KV^n\|,$$

whence $\|H(a)\|_{ess} \geqq \operatorname{dist}_{L^\infty}(\overline{a}, C + H^\infty)$. Δ

2.17. Before proceeding to Toeplitz operators with symbols from $C + H^\infty$, it seems us to be convenient to make some remarks concerning compactness of Hankel operators. For the proofs we refer to the relevant literature.

In 2.16, we saw that $H(a) \in \mathfrak{C}_\infty(H^2)$ if and only if $a \in C + \overline{H^\infty}$.
In this connection note the following results.

(a) (Kronecker) $H(a)$ is of finite rank if and only if $a \in R + \overline{H^\infty}$, where R denotes the set of rational functions without poles on Γ.

(b) (Peller) $H(a) \in \mathfrak{C}_p(H^2)$ $(1 \leqq p < \infty)$ if and only if $a \in B_p^{1/p} + \overline{H^\infty}$ (see 1.20).

(c) (Adamian, Arov, Krein) Let $a \in L^\infty$. According to 1.2, $s_n(H(a))$ is equal to

$$s_n(H(a)) = \inf\{\|H(a) - K\| : \dim \operatorname{Im} K \leqq n\}.$$

It turns out that this infimum can be taken over the collection of all Hankel operators of rank at most n:

$$s_n(H(a)) = \inf\{\|H(a) - H(f)\| : \dim \operatorname{Im} H(f) \leqq n\}.$$

2.18. Theorem (Douglas). Let $a \in C + H^\infty$. Then

$$T(a) \in \Phi(H^2) \iff a \in G(C + H^\infty).$$

Proof. Assume $a,b \in C + H^\infty$, $ab = \chi_0$. Then $H(\tilde{a}), H(\tilde{b}) \in \mathfrak{C}_\infty(H^2)$ by Theorem 2.16, and from Proposition 2.7 we get

$$T(a)T(b) - I = - H(a)H(\tilde{b}) \in \mathfrak{C}_\infty(H^2),$$
$$T(b)T(a) - I = - H(b)H(\tilde{a}) \in \mathfrak{C}_\infty(H^2),$$

whence $T(a) \in \Phi(H^2)$ by 1.12(a).
Now assume $T(a) \in \Phi(H^2)$. Theorem 2.10 gives the invertibility of $a(t)$ in L^∞, and then the lemma in 1.18 yields the existence of a unimodular function $u \in C + H^\infty$ and an outer function $h \in H^\infty$ such that $a = uh$. By the theorem in 1.18, $h \in GH^\infty$. Thus $T(h^{-1}) \in \Phi(H^2)$ and combining this with our assumption that $T(a) \in \Phi(H^2)$, we arrive at $T(u) = T(a)T(h^{-1}) \in \Phi(H^2)$. Let n denote the index of $T(u)$. Then $\operatorname{Ind} T(u\chi_n) = 0$, and by the corollary in 2.8, $T(u\chi_n)$ is invertible. This implies the invertibility of $T(\bar{u}\chi_{-n}) = T^*(u\chi_n)$. Due to Theorem 2.15 there is an outer function $f \in H^\infty$ satisfying

$$\|\bar{u}\, \chi_{-n} - f\|_{L^\infty} = \|1 - \chi_n u f\|_{L^\infty} < 1.$$

Hence $\chi_n uf \in G(C + H^\infty)$ and consequently, $u \in G(C + H^\infty)$, which, finally, implies that $a \in G(C + H^\infty)$. Δ

Remark 1. Recall the invertibility criterion for $C + H^\infty$ given in 1.19. The index of a Fredholm Toeplitz operator with symbol from $C + H^\infty$ can be shown to be

$$\operatorname{Ind} T(a) = - \operatorname{ind} a.$$

Remark 2. Another proof of this theorem will be given in 4.5.

CONTINUOUS SYMBOLS

2.19. The classes $M\langle p\rangle$. For $p \in (1,2) \cup (2,\infty)$ let $M\langle p\rangle$ denote the collection of all functions $a \in L^\infty$ belonging to $M_{\tilde{p}}$ for all $\tilde{p}$ in some neighborhood of p. Furthermore, put $M_{\langle 1\rangle} = M_1$ and $M_{\langle 2\rangle} = M_2$.

2.20. Proposition. If $a \in C \cap M\langle p\rangle$ then $H(a) \in \mathfrak{C}_\infty(l^p)$.

Proof. If $p = 1$ then

$$\|H(a) - H(P_n Pa)\|_{\mathfrak{B}(l^1)} = \|H(Q_n Pa)\|_{\mathfrak{B}(l^1)} \leqq \|Q_n Pa\|_W =$$

$$= \sum_{k=n+1}^{\infty} |a_k| = o(1) \quad \text{as} \quad n \to \infty$$

and the compactness of $H(a)$ follows from 1.1(e). The case $p = 2$ is covered by Theorem 2.16. Thus let $p \in (1,2) \cup (2,\infty)$. By our hypothesis we can choose a sufficiently small ε, $\varepsilon < 0$ for $1 < p < 2$ and $\varepsilon > 0$ for $2 < p < \infty$, such that $a \in C \cap M_{p+\varepsilon}$. Then there is a $\gamma \in (0,1)$ satisfying

$$1/p = \gamma/(p + \varepsilon) + (1 - \gamma)/2$$

and applying the Riesz-Thorin interpolation theorem we get

$$\|Q_n H(a)\|_{\mathcal{B}(l^p)} \leq \|Q_n H(a)\|^{\gamma}_{\mathcal{B}(l^{p+\varepsilon})} \|Q_n H(a)\|^{1-\gamma}_{\mathcal{B}(l^2)} .$$

Due to the estimate $\|Q_n H(a)\| \leq \|H(a)\|$ the first factor on the right remains bounded, whereas the second converges to zero by Theorem 2.16 and 1.3(d). Hence

$$\|H(a) - P_n H(a)\|_{\mathcal{B}(l^p)} = \|Q_n H(a)\|_{\mathcal{B}(l^p)} = o(1) \quad \text{as} \quad n \to \infty,$$

and application of 1.1(e) completes the proof. Δ

<u>2.21. Theorem.</u> Let $a \in C \cap M_{<p>}$ $(1 \leq p < \infty)$. Then $T(a) \in \Phi(l^p)$ if and only if $a(t) \neq 0$ for every $t \in \Gamma$. In that case $\operatorname{Ind} T(a) = - \operatorname{ind} a$, the function a^{-1} belongs to $C \cap M_{<p>}$, and $T(a^{-1})$ is a regularizer for $T(a)$.
In particular, $T(a) \in G\mathcal{B}(l^p)$ if and only if $a(t)$ does not vanish on Γ and has winding number zero.

<u>Proof.</u> If $T(a) \in \Phi(l^p)$ then $a(t) \neq 0$ for every $t \in \Gamma$ by Theorem 2.10. Suppose now $a(t)$ does not vanish on Γ.
First of all we shall prove that then $a^{-1} \in M_{<p>}$. In case $p = 1$ this is a consequence of the well-known Wiener-Levy theorem (see e.g. Rudin [2], 11.6).
For $p > 1$ we apply the local principle of Gohberg and Krupnik. To that end let the $\mathcal{A}$ in 1.14 be the (commutative) Banach algebra M_p (see 2.2).
Given $\tau = e^{i\vartheta} \in \Gamma$ put

$$\begin{aligned} N_\tau = \{f \in C : 0 \leq f \leq 1, &\text{ there is an } \varepsilon > 0 \text{ such that} \\ &f(e^{i\varphi}) \equiv 1 \text{ for } |\varphi - \vartheta| < \varepsilon, \\ &f \text{ is monotonically increasing for } \vartheta - 2\varepsilon < \varphi < \vartheta - \varepsilon, \\ &f \text{ is monotonically decreasing for } \vartheta + \varepsilon < \varphi < \vartheta + 2\varepsilon, \\ &f(e^{i\varphi}) \equiv 0 \text{ for } |\varphi - \vartheta| > 2\varepsilon\}. \end{aligned}$$

Note that the total variation of each function in N_τ equals 2, so $N_\tau \subset M_p$ by 2.2(5). It is easily shown that $\{N_\tau\}_{\tau\in\Gamma}$ forms a covering system of localizing classes in M_p. Applying the Riesz-Thorin interpolation theorem in a similar fashion as in the preceding proof one can show that $a \in C \cap M_{<p>} \subset M_p$ is N_τ-equivalent to a_τ given by $a_\tau(t) \equiv a(\tau)$ $(t \in \Gamma)$. Obviously, $a_\tau \in GM_p$. The theorem in 1.14 then gives $a^{-1} \in M_p$. Repeating this argument for all $\tilde{p}$ in some neighborhood of p, we arrive at the desired relation $a^{-1} \in M_{<p>}$.

Thus $T(a^{-1}) \in \mathfrak{B}(l^p)$. By virtue of Proposition 2.20 it remains to apply Proposition 2.7 in order to conclude that $T(a^{-1})$ is a regularizer of $T(a)$.

Finally, the equality $\operatorname{Ind} T(a) = -\operatorname{ind} a$ can be proved by a simple homotopy argument. Δ

Remark 1. Note that the preceding theorem gives a full information about the spectra of $T(a)$ in $\mathfrak{B}(l^p)$ if $a \in C \cap M_{<p>}$ $(1 \leqq p < \infty)$:

$$\sigma_{ess}(T(a)) = R(a),$$
$$\sigma(T(a)) = R(a) \cup \{\lambda \in \mathbb{C} : i(a,\lambda) \neq 0\},$$

where $R(a)$ refers to the range of $a(t)$ and where $i(a,\lambda)$ denotes the winding number of the oriented curve $(t \in \Gamma) \mapsto (a(t) \in \mathbb{C})$ with respect to λ.

Remark 2. For $a \in C \cap M_{<p>}$ the following is true: $T(a) \in \Phi_+(l^p)$ or $T(a) \in \Phi_-(l^p)$ implies $T(a) \in \Phi(l^p)$.
In fact, suppose $T(a) \in \Phi_+(l^p)$ and $\operatorname{Ind} T(a) = -\infty$. One can show that for any $\varepsilon > 0$ there is a $b \in C \cap M_{<p>}$ such that $\|T(a) - T(b)\|_{\mathfrak{B}(l^p)} < \varepsilon$ and $b(t) \neq 0$ for every $t \in \Gamma$. By the above theorem $|\operatorname{Ind} T(b)| < \infty$, whereas according to 1.12(d),(f) $\operatorname{Ind} T(b) = -\infty$. This is a contradiction. Consequently, $T(a) \in \Phi(l^p)$.

PIECEWISE CONTINUOUS SYMBOLS

2.22. The algebras PC and $\overline{PC}$. Let PC denote the algebra of all functions $a \in L^\infty$ that are continuous with exception of finitely many points $t_1,\dots,t_R \in \Gamma$, where, however, the finite limits $a(t_r - 0)$ and $a(t_r + 0)$ exist $(r = 1,\dots,R)$. We denote by $\overline{PC}$ the closure of PC in L^∞. Note that a function in $\overline{PC}$ has at most countably many discontinuities of first kind (jumps). For our purposes there is no loss of generality in assuming that all functions $a \in \overline{PC}$ are continuous from the left, i.e. that $a(t) = a(t - 0)$ for every $t \in \Gamma$ holds.

Given $a \in \overline{PC}$ we denote by $R_2(a)$ the continuous and oriented curve obtained from the range of $a(t)$ by filling in the straight line segments joining $a(t-0)$ to $a(t+0)$ for each discontinuity. Thus $R_2(a)$ is just the image of the cylinder $\mathfrak{Z} = \Gamma \times [0,1]$ under the mapping defined by

$$(\zeta,\mu) \in \Gamma \times [0,1] \mapsto (1-\mu)a(\zeta-0) + \mu a(\zeta+0).$$

<u>2.23. An identity.</u>

$$T(af^2b) - T(a)T(f^2)T(b)$$

$$= T(af \cdot fb) - T(af)T(fb)$$

$$+ T(af)H(f)H(\tilde{b}) + H(a)H(f)T(fb)$$

$$- H(a)H(\tilde{f})H(f)H(\tilde{b}) - T(a)H(f)H(\tilde{f})T(b).$$

<u>Proof.</u> By Proposition 2.7,

$$T(af^2b) - T(a)T(f^2)T(b)$$

$$= T(af \cdot fb) - T(a)T(f)T(f)T(b) - T(a)H(f)H(\tilde{f})T(b)$$

$$= T(af \cdot fb) - [T(af) - H(a)H(\tilde{f})] \cdot [T(fb) - H(f)H(\tilde{b})]$$

$$- T(a)H(f)H(\tilde{f})T(b). \ \Delta$$

<u>2.24. Proposition.</u> Let $a,b \in M_{<p>}$ $(1 \leq p < \infty)$ and suppose to each $\tau \in \Gamma$ there is a neighborhood $U_\tau \subset \Gamma$ such that $a|U_\tau$ or $b|U_\tau$ is continuous. Then $T(ab) - T(a)T(b) \in \mathfrak{C}_\infty(l^p)$.

<u>Proof.</u> Since Γ is compact, we can choose a finite collection $\{U_{\tau_i}\}$ of such neighborhoods forming a covering of Γ. Let $\sum g_i = 1$ be a subordinate smooth partition of unity (cf. Rudin [2], 6.20). Obviously, each g_i can be represented as $g_i = f_i^2$ where f_i is also smooth. Then

$$T(ab) - T(a)T(b) = \sum[T(af_i^2b) - T(a)T(f_i^2)T(b)].$$

But $H(f_i)$, $H(\tilde{f}_i) \in \mathfrak{C}_\infty(l^p)$ by Proposition 2.20, and

$$T(af_i \cdot f_ib) - T(af_i)T(f_ib) = H(af_i)H(\tilde{f}_i\tilde{b}) \in \mathfrak{C}_\infty(l^p)$$

since by our hypotheses at least one of the functions af_i and f_ib is continuous and so Proposition 2.20 again applies. Now the assertion immediately follows from the identity 2.23. Δ

<u>Corollary.</u> If $a,b \in \overline{PC} \cap M_{<p>}$ have no common points of discontinuity, then $T(ab) - T(a)T(b) \in \mathfrak{C}_\infty(l^p)$.

2.25. Proposition. If $a,b \in \overline{PC} \cap M_{<p>}$ then

$$T(a)T(b) - T(b)T(a) \in \mathfrak{C}_\infty(l^p).$$

Proof. A little thought shows that it suffices to prove the assertion for the case where $a(t)$ and $b(t)$ have only one point of discontinuity α and β, respectively. If $\alpha \neq \beta$ then write

$$T(a)T(b) - T(b)T(a) = [T(ba)-T(b)T(a)] - [T(ab) - T(a)T(b)]$$

and apply the corollary in 2.24. Thus let $\alpha = \beta$. Then there exist a constant $\lambda \in \mathbb{C}$ and a function $c \in C \cap M_{<p>}$ such that $a = \lambda b + c$. Hence

$$T(a)T(b) - T(b)T(a) = T(\lambda b + c)T(b) - T(b)T(\lambda b + c) =$$

$$= T(c)T(b) - T(b)T(c) = [T(bc)-T(b)T(c)] - [T(cb) - T(c)T(b)]$$

and the assertion again follows from the corollary in 2.24. Δ

2.26. Proposition. Let $a,b \in L^\infty$ and $1 \leqq p \leqq \infty$. Then the following two statements are equivalent:

(a) $T(ab) - T(a)T(b) \in \mathfrak{C}_p(l^2)$.

(b) To each $\tau \in \Gamma$ there exist a neighborhood $U_\tau \subset \Gamma$ and two functions $a_\tau, b_\tau \in L^\infty$ such that

$$a|U_\tau = a_\tau|U_\tau, \quad b|U_\tau = b_\tau|U_\tau, \quad T(a_\tau b_\tau)-T(a_\tau)T(b_\tau) \in \mathfrak{C}_p(l^2).$$

Proof. The implication (a) ==> (b) is trivial. Let us prove that (b) implies (a).
From the compactness of Γ we deduce the existence of a finite set $\{U_{\tau_i}\}$ of neighborhoods satisfying (b) such that $\Gamma \subset \cup U_{\tau_i}$. Denote the corresponding a_{τ_i} and b_{τ_i} by a_i and b_i. Now let $\sum f_i^2 = 1$ be a subordinate smooth partition of unity (cf. the proof of Proposition 2.24). Then

$$T(ab) - T(a)T(b) = \sum[T(af_i^2b) - T(a)T(f_i^2)T(b)].$$

The identity 2.23 shows that each item of this sum is modulo $\mathfrak{C}_p(l^2)$ equal to

$$T(af_i \cdot f_ib) - T(af_i)T(f_ib) = T(a_if_i \cdot f_ib_i)-T(a_if_i)T(f_ib_i). \quad (11)$$

But a few application of Proposition 2.7 gives (we drop the subscript i)

$$\begin{aligned} & T(af \cdot fb) - T(af)T(fb) \\ = \; & T(fa \cdot bf) - T(fa)T(bf) = H(fa)H(\widetilde{bf}) \\ = \; & [T(f)H(a) + H(f)T(\tilde{a})] \cdot [T(\tilde{b})H(\tilde{f}) + H(\tilde{b})T(f)] \end{aligned}$$

$$= T(f)[T(ab) - T(a)T(b)]T(f) + T(f)H(a)T(\tilde{b})H(\tilde{f}) +$$
$$+ H(f)H(\tilde{a})T(\tilde{b})H(\tilde{f}) + H(f)T(\tilde{a})H(\tilde{b})T(f)$$

which by our hypotheses immediately implies that (11) belongs to $\mathfrak{C}_p(l^2)$. Δ

Corollary 1. Let $a,b \in L^\infty$ and $1 \leq r < \infty$. Furthermore, suppose each $\tau \in \Gamma$ has a neighborhood U_τ such that either

$a(t)$ belongs to $B_{p_\tau}^{1/p_\tau}$ and $b(t)$ to $B_{q_\tau}^{1/q_\tau}$ in U_τ, where $1 < p_\tau < \infty$, $1/p_\tau + 1/q_\tau = 1/r$

or

one of the functions $a(t)$ and $b(t)$ belongs to B_1^1 in U_τ.

Then $T(ab) - T(a)T(b) \in \mathfrak{C}_r(l^2)$.

To see this recall 2.17(b).

Corollary 2. Let $a,b \in \overline{PC}$. Suppose for every $\tau \in \Gamma$ the following conditions are fulfilled:

if both $a(t)$ and $b(t)$ are continuous at τ, then there is a neighborhood of τ in which both $a(t)$ and $b(t)$ belong to $B_2^{1/2}$; if one of these functions has a discontinuity at τ, then the other belongs to B_1^1 in some neighborhood of τ.

Under this hypotheses $T(ab) - T(a)T(b) \in \mathfrak{C}_1(l^2)$.

2.27. In 2.17, we listed some results concerning compactness of Hankel operators. But these questions are of (perhaps, even greater) interest for products of Hankel operators, too. Since

$$H(a)H(\tilde{b}) = T(ab) - T(a)T(b),$$

the problems are usually posed and solved in the language of quasi-commutators $T(ab) - T(a)T(b)$. Certain questions are answered by the Propositions 2.24 and 2.26. The following facts represent some much deeper results along these lines.

For $a \in L^\infty$ denote by $\mathrm{alg}(H^\infty,a)$ the smallest closed subalgebra of L^∞ containing H^∞ and $a(t)$.

(a) (Brown, Halmos) If $a,b \in L^\infty$ then

$$T(ab) - T(a)T(b) = 0 \iff a \in H^\infty \text{ or } b \in H^\infty$$
$$\iff \mathrm{alg}(H^\infty,\bar{a}) \cap \mathrm{alg}(H^\infty,b) = H^\infty.$$

(b) (Axler, Chang, Sarason, Vol'berg) If $a,b \in L^\infty$ then

$$T(ab)-T(a)T(b) \in \mathfrak{C}_\infty(l^2) \iff \mathrm{alg}(H^\infty,\bar{a}) \cap \mathrm{alg}(H^\infty,b) \subset C + H^\infty.$$

In this connection recall the first theorem of 1.19.

(c) (Axler, Chang, Sarason) If $a,b \in L^\infty$ then

$$T(ab)-T(a)T(b) \in \mathfrak{C}_0(l^2) \Longleftrightarrow H(a) \in \mathfrak{C}_0(l^2) \quad \text{or} \quad H(\tilde{b}) \in \mathfrak{C}_0(l^2).$$

2.28. A representation of functions from PC. For $\tau \in \Gamma$ and $\alpha \in \mathbb{C}$ we define the function φ_τ^α by

$$\varphi_\tau^\alpha(t) = \exp\{\alpha i \arg(-t/\tau)\} \quad (t \in \Gamma), \tag{12}$$

where, by convention, $|\arg(-t/\tau)| < \pi$. So φ_τ^α is a branch of the function $(-t)^\alpha$, having exactly one point of discontinuity, namely, τ. Let now $a \in PC$. Denote the points of discontinuity by $t_1,\ldots,t_R$. A simple geometrical consideration shows that in the case where $0 \notin R_2(a)$, to each $r = 1,\ldots,R$ a number $\beta_r \in \mathbb{C}$ can be found which satisfies

$$a(t_r - 0)/a(t_r + 0) = \exp(2\pi i\beta_r)$$

and

$$-1/2 < \operatorname{Re} \beta_r < 1/2.$$

Then it is easily shown that there is a $b \in C$ such that $a(t)$ admits the representation

$$a = \varphi_{t_1}^{\beta_1} \cdots \varphi_{t_R}^{\beta_R} b\,. \tag{13}$$

2.29. Theorem. Let $a \in PC$. Then $T(a) \in \Phi(H^2)$ if and only if $0 \notin R_2(a)$. In that case

$$\operatorname{Ind} T(a) = -\operatorname{ind} R_2(a). \tag{14}$$

Proof. Suppose $0 \notin R_2(a)$. Then $a(t)$ has the representation 2.28(13) and because of the corollary in 2.24,

$$T(a) - T(\varphi_{t_1}^{\beta_1}) \cdots T(\varphi_{t_R}^{\beta_R})\, T(b) \in \mathfrak{C}_\infty(H^2). \tag{15}$$

Since b does not vanish on Γ, we have $T(b) \in \Phi(H^2)$ by Theorem 2.21. Furthermore, $R_2(\varphi_{t_r}^{\beta_r})$ is completely contained in the interior of a sector spanned by an angle less than π and having its vertex at the origin. By Theorem 2.11, $T(\varphi_{t_r}^{\beta_r}) \in G\mathfrak{B}(H^2)$. Thus $T(a) \in \Phi(H^2)$.

Let us now prove the formula (14). Because of (15) and the invertibility of $T(\varphi_{t_r}^{\beta_r})$, we have $\operatorname{Ind} T(a) = \operatorname{Ind} T(b)$. Due to Theorem 2.21, $\operatorname{Ind} T(b) = -\operatorname{ind} b = -\operatorname{ind} R_2(b)$. It remains to verify the equality $\operatorname{ind} R_2(b) = \operatorname{ind} R_2(a)$. But this follows from the fact that $\operatorname{ind} R_2(fg) = \operatorname{ind} R_2(f) + \operatorname{ind} R_2(g)$, whenever $f,g \in PC$ have no common points of discontinuity.

Finally, suppose $0 \in R_2(a)$ but $T(a) \in \Phi(H^2)$. There exist small perturbations b and c of $a(t)$ such that $|\text{ind } R_2(b) - \text{ind } R_2(c)| = 1$. On the other hand, by virtue of 1.12(d), $\text{Ind } T(b) = \text{Ind } T(c) = \text{Ind } T(a)$, which by the already proved formula (14) leads to $\text{ind } R_2(b) - \text{ind } R_2(c) = 0$. So we have a contradiction, i.e. $T(a) \in \Phi(H^2)$ actually implies that $0 \notin R_2(a)$. Δ

__2.30. Theorem.__ Theorem 2.29 remains true for $a \in \overline{PC}$.

__Proof.__ We shall apply the local principle 1.14. For this purpose put $\mathcal{A} = \mathcal{B}(H^2)/\mathcal{C}_\infty(H^2)$ and denote the canonical projection of $\mathcal{B}(H^2)$ onto $\mathcal{A}$ by π (cf. 1.13). Next, for $\tau \in \Gamma$ define

$$N_\tau = \{f \in C : 0 \leqq f \leqq 1,\ \ f \equiv 1 \text{ in some neighborhood of } \tau\},$$
$$M_\tau = \{\pi\, T(f) \in \mathcal{A} : f \in N_\tau\}.$$

It is easy to show that $\{M_\tau\}_{\tau\in\Gamma}$ forms a covering system of localizing classes in $\mathcal{A}$ and that $\pi T(a)$ commutes with $\pi T(f)$, $f \in N_\tau$ (the latter fact follows from the corollary in 2.24).
Suppose $0 \notin R_2(a)$. A simple computation gives $\pi T(a) \overset{M_\tau}{\sim} \pi T(a_\tau)$, where a_τ is defined by

$$a_\tau(t) = \begin{cases} a(\tau - 0), & -\pi < \arg(t/\tau) < 0 \\ a(\tau + 0), & 0 < \arg(t/\tau) < \pi \end{cases}.$$

Now, $\text{conv } R(a_\tau)$ is just the straight line segment joining $a(\tau - 0)$ to $a(\tau + 0)$, and if $0 \notin R_2(a)$, then with the help of Theorem 2.11 we may conclude that $\pi T(a_\tau) \in G\mathcal{A}$. To show that $T(a) \in \Phi(H^2)$, it remains to apply the theorem from 1.14.

The index formula will be our next concern. By 1.12(d) there is an $\varepsilon > 0$ such that $T(b) \in \Phi(H^2)$ for all $b \in \overline{PC}$ satisfying $\|b - a\|_{L^\infty} < \varepsilon$. Thereby $\text{Ind } T(b) = \text{Ind } T(a)$. Choosing such a function b from PC we have, by Theorem 2.24, $\text{Ind } T(b) = -\text{ind } R_2(b)$. From $\text{ind } R_2(b) = \text{ind } R_2(a)$ it follows that $\text{Ind } T(a) = -\text{ind } R_2(a)$.

Finally, assume $0 \in R_2(a)$ but $T(a) \in \Phi(H^2)$. Then there exists a function $b \in PC$ having the following properties: $0 \in R_2(b)$, $T(b) \in \Phi(H^2)$, $\text{Ind } T(b) = \text{Ind } T(a)$. Now by Theorem 2.29 we can find two small perturbations $c, d \in PC$ of b such that

$$T(c), T(d) \in \Phi(H^2), \quad |\text{Ind } T(c) - \text{Ind } T(d)| = 1,$$

whereas in view of 1.12(d)

$$\text{Ind } T(c) = \text{Ind } T(d) = \text{Ind } T(b).$$

Thus we have a contradiction. It follows that $0 \notin R_2(a)$ if $T(a) \in \Phi(H^2)$.
Δ

2.31. The case $p \neq 2$. For $p \neq 2$ put $\vartheta = \pi(1 - \frac{2}{p})$ and define for $\mu \in [0,1]$

$$s_p(\mu) = \frac{\sin \vartheta\mu \exp(i\vartheta\mu)}{\sin \vartheta \exp(i\vartheta)} .$$

If μ runs from 0 to 1, then $s_p(\mu)$ runs in $\mathbb{C}$ for $1 < p < 2$ (resp. $2 < p < \infty$) along a circular arc joining 0 to 1 and located at the left (resp. right) of the line segment $[0,1]$.
For $a \in \overline{PC}$ we define $R_p(a)$ to be the continuous, closed, oriented curve obtained from the range of $a(t)$ by filling in the arcs

$$\{(1 - s_p(\mu))a(t - 0) + s_p(\mu)a(t + 0) : \mu \in [0,1]\}$$

for each discontinuity.

Theorem. Suppose $a \in \overline{PC} \cap M_{<p>}$, $1 < p < \infty$. Then $T(a) \in \Phi(l^p)$ if and only if $0 \notin R_p(a)$. In that case

$$\operatorname{Ind} T(a) = - \operatorname{ind} R_p(a). \qquad (16)$$

Proof outline.
Step 1. Let φ_τ^α be the function defined by 2.28(12). It can be proved (Dudučava [3]) that $T(\varphi_\tau^\alpha)$ is invertible on l^p if $-1/p < \operatorname{Re} \alpha < 1/q$, where $1/p + 1/q = 1$, and that its inverse is $cLT[\varphi_\tau^{(-\alpha)}]M$, where $c = \pi\alpha/\sin \pi\alpha$, $L = \operatorname{diag}(\lambda_j)_{j=0}^\infty$, $M = \operatorname{diag}(\mu_j)_{j=0}^\infty$, $\lambda_0 = \mu_0 = 1$,

$$\lambda_j = \prod_{k=1}^{j} (1 + \frac{\alpha}{k}), \quad \mu_j = \prod_{k=1}^{j} (1 - \frac{\alpha}{k}) \quad (j \geq 1).$$

Note that

$$-1/p < \operatorname{Re} \alpha < 1/q \iff 0 \notin R_p(\varphi_\tau^\alpha) \text{ and } \operatorname{ind} R_p(\varphi_\tau^\alpha) = 0.$$

Step 2. If $a \in PC$ then $a(t)$ admits the representation 2.28(13). From the corollary in 2.24 we deduce

$$T(a) - T(\varphi_{t_1}^{\beta_1}) \cdots T(\varphi_{t_R}^{\beta_R})\, T(b) \in \mathfrak{C}_\infty(l^p).$$

Together with step 1 this gives that $T(a) \in \Phi(l^p)$ if only $a \in PC \cap M_{<p>}$ and $0 \notin R_p(a)$.

Step 3. The index formula (16) for $a \in PC \cap M_{<p>}$ may be verified in the same way as for $p = 2$.

Step 4. The local principle 1.14 and step 2 can be used to show that $T(a) \in \Phi(l^p)$ for every $a \in \overline{PC} \cap M_{<p>}$ such that $0 \notin R_p(a)$ (confer the proof of Theorem 3.19 for the technical details of such a localization).

Step 5. The index formula (16) for $a \in \overline{PC} \cap M_{<p>}$ results from steps 3 and 4 through an approximation argument (cf. the proof of Theorem 2.30).

Step 6. Once (16) is proved, the necessity part of our theorem follows by the usual perturbation argument (see the proof of Theorem 2.30, too). Δ

Remark 1. Combining the preceding theorem with the corollary in 2.8 we get for $a \in \overline{PC} \cap M_{<p>}$ $(1 < p < \infty)$

$$T(a) \in G\mathcal{B}(l^p) \iff 0 \notin R_p(a) \text{ and } \operatorname{ind} R_p(a) = 0.$$

Remark 2. The condition that $0 \notin R_p(a)$ is even necessary (and sufficient) for $T(a)$, where $a \in \overline{PC} \cap M_{<p>}$ $(1< p < \infty)$, to be a Φ_+- or Φ_--operator (cf. Remark 2 in 2.21).

MATRIX SYMBOLS

2.32. Block Toeplitz operators. A system of Toeplitz operators in l^p may be regarded as an operator matrix $(T(a_{jk}))_{j,k=1}^N$ acting on l_N^p (recall the notations of 1.15).

Let $\mathbb{C}_N \otimes l^p$ denote the $\mathbb{C}_N$-valued l^p-space and let $F_N : l_N^p \to \mathbb{C}_N \otimes l^p$ be the mapping which sends

$$\varphi = (\{\varphi_i^{(1)}\}_{i=1}^\infty, \dots, \{\varphi_i^{(N)}\}_{i=1}^\infty)^\tau \in l_N^p$$

to

$$\psi = \{\psi_1, \psi_2, \psi_3, \dots\}, \quad \psi_i = (\varphi_i^{(1)}, \dots, \varphi_i^{(N)})^\tau \in \mathbb{C}_N.$$

Equipped with the norm

$$\|\psi\|_{\mathbb{C}_N \otimes l^p} = \Big(\sum_{i \in \mathbb{Z}_+} (|\varphi_i^{(1)}|^p + \dots + |\varphi_i^{(N)}|^p)\Big)^{1/p}$$

$\mathbb{C}_N \otimes l^p$ becomes a Banach space. It is easily seen that F_N is a homeomorphical isomorphism of the Banach spaces l_N^p and $\mathbb{C}_N \otimes l^p$. Now the operator $F_N(T(a_{jk}))_{j,k=1}^N F_N^{-1} \in \mathcal{B}(\mathbb{C}_N \otimes l^p)$ may be written as $T(a) = (a_{l-m})_{l,m=0}^\infty$ where a_i $(i \in \mathbb{Z})$ denotes the $N \times N$-matrix $a_i = ((a_{jk})_i)_{j,k=1}^N$ formed by the Fourier coefficients of a_{jk}. For a natural reason this operator $T(a)$ is called block Toeplitz operator.

In what follows we shall identify l_N^p with $\mathbb{C}_N \otimes l^p$, and we shall speak henceforth about " the block Toeplitz operator $T(a)$ on l_N^p ". In each case it is seen from the context which operator we shall have in mind, the system $(T(a_{jk}))_{j,k=1}^N$ on l_N^p or the operator

$T(a)$, $a = (a_{jk})_{j,k=1}^{N}$ on $\mathbb{C}_N \otimes l^p$.

The matrix-function given by

$$a : \Gamma \to \mathbb{C}_{N\times N}, \quad t \mapsto (a_{jk}(t))_{j,k=1}^{N}$$

is called the <u>symbol</u> of the block Toeplitz operator $T(a)$ on l^p_N.
Thinking of $\{\varphi_j\}_{j\in\mathbb{Z}}$ in 2.1(1) and 2.3(6) as an element of $\mathbb{C}_N \otimes l^p$, we get the definition of $M(a)$ and $H(a)$ for $a = (a_{jk})_{j,k=1}^{N}$.
Furthermore, let P and Q denote not only the natural projections on Υ^p (see 2.4) but also the corresponding projections on $\tilde{l}^p_N$.
In particular, Proposition 2.7 remains true for the matrix case.

<u>2.33. Theorem.</u> Let $a \in (M_p)_{N\times N}$, $1 \leqq p < \infty$. If $T(a) \in \Phi(l^p_N)$ then $a \in G(M_p)_{N\times N}$ and, consequently, $a \in GL^{\infty}_{N\times N}$.

For $p > 1$ the arguments of the proof of Theorem 2.10 apply if one only sets $U = M(\chi_1 I_{N\times N})$, where $I_{N\times N}$ is the identity matrix of order n. For $p = 1$ the problem can be reduced to the scalar case by means of the corollary in 1.15.

<u>2.34. Proposition.</u> Let $a \in GL^{\infty}_{N\times N}$. Then

$$T(a) \in G\mathcal{B}(H^2_N) \iff T(\overline{a}) \in G\mathcal{B}(H^2_N) \tag{17}$$

and

$$T(\tilde{a}) \in G\mathcal{B}(H^2_N) \iff T(a^{-1}) \in G\mathcal{B}(H^2_N). \tag{18}$$

If $N = 1$ then in addition

$$T(a) \in G\mathcal{B}(H^2) \iff T(\tilde{a}) \in G\mathcal{B}(H^2), \tag{19}$$

but for $N > 1$ the invertibility of $T(a)$ on H^2_N is, in general, <u>not</u> equivalent to that of $T(\tilde{a})$ on H^2_N.

<u>Comments.</u> Here, as usual, the matrix-functions $\tilde{a}$ and $\overline{a}$ are defined by

$$\tilde{a}(t) = a(1/t), \quad \overline{a}(t) = \overline{a(t)^{\tau}} \quad (t \in \Gamma),$$

where τ denotes transponation. Furthermore, note that by Theorem 2.33 the invertibility of $a \in L^{\infty}_{N\times N}$ in $L^{\infty}_{N\times N}$ is a consequence of the invertibility of $T(a)$ (or $T(\tilde{a})$, or $T(\overline{a})$) on H^2_N.

<u>Proof.</u> The equivalence (17) is obvious, since $T(\overline{a}) = T^*(a)$.
If $T(\tilde{a}) \in G\mathcal{B}(H^2_N)$ then through an easy computation can be shown that

$$T(a) - H(a)T^{-1}(\tilde{a})H(\tilde{a})$$

is the inverse of $T(a^{-1})$, and if $T(a^{-1}) \in G\mathcal{B}(H^2_N)$ then

$$T(\widetilde{a}^{-1}) - H(\widetilde{a}^{-1})T^{-1}(a^{-1})H(a^{-1})$$

can be shown to be the inverse of $T(\widetilde{a})$. Thus (18) holds, too. Next, the equivalence (19) follows from (17) and $T(\widetilde{a}) = WT(\overline{a})W$, where W denotes the isometry on H^2 defined by $(W\varphi)(t) = \overline{\varphi(1/t)}$.

Finally, put

$$a(t) = \begin{pmatrix} t & 1 \\ 0 & t^{-1} \end{pmatrix}.$$

Then $a = h_-h_+$ and $\widetilde{a} = g_-dg_+$, where

$$h_-(t) = \begin{pmatrix} 1 & 0 \\ t^{-1} & 1 \end{pmatrix}, \quad h_+(t) = \begin{pmatrix} t & 1 \\ -1 & 0 \end{pmatrix},$$

$$g_-(t) = \begin{pmatrix} t^{-1} & 1 \\ 1 & 0 \end{pmatrix}, \quad d(t) = \begin{pmatrix} t & 0 \\ 0 & t^{-1} \end{pmatrix}, \quad g_+(t) = \begin{pmatrix} 0 & 1 \\ 1 & 0 \end{pmatrix}.$$

We have $h_-, g_- \in \overline{GH^\infty_{2\times 2}}$ and $h_+, g_+ \in GH^\infty_{2\times 2}$, consequently, by Proposition 2.7, $T(a) \in G\mathcal{B}(H^2_2)$ and

$$\dim \operatorname{Ker} T(\widetilde{a}) = \dim \operatorname{Ker} T(d) = 1,$$

$$\dim \operatorname{Coker} T(\widetilde{a}) = \dim \operatorname{Coker} T(d) = 1,$$

i.e. $T(\widetilde{a}) \notin G\mathcal{B}(H^2_2)$. This completes the proof. Δ

<u>Remark 1.</u> It should be noted here that in view of $T(\widetilde{a}) = JQM(a)QJ|H^2$ and $T(a^{-1}) = PM^{-1}(a)P|H^2$ the equivalence (18) is an immediate consequence of Proposition 3.8 below.

<u>Remark 2.</u> In general Theorem 2.8 and its corollary are <u>not</u> valid in the matrix case.
Example: if $a(t) = \operatorname{diag}(t^{\varkappa_1}, \dots, t^{\varkappa_N})$ then

$$\dim \operatorname{Ker} T(a) = -\sum_{\varkappa_j<0} \varkappa_j ,$$

$$\dim \operatorname{Coker} T(a) = \sum_{\varkappa_j>0} \varkappa_j ;$$

in particular, $N = 2$, $\varkappa_1 = 1$, $\varkappa_2 = -1$ gives a counterexample.

<u>Remark 3.</u> For $a \in GL^\infty_{N\times N}$ we have

$$T(a) \in \Phi(H^2_N) \iff T(\overline{a}) \in \Phi(H^2_N)$$

and

$$T(\widetilde{a}) \in \Phi(H^2_N) \iff T(a^{-1}) \in \Phi(H^2_N).$$

The first equivalence is obvious. Clearly, $\operatorname{Ind} T(a) = -\operatorname{Ind} T(\overline{a})$.

To verify the second equivalence notice that if R and S are regularizers of $T(\tilde{a})$ and $T(a^{-1})$, respectively, then

$$T(a) - H(a)RH(\tilde{a}) \quad \text{and} \quad T(\widetilde{a^{-1}}) - H(\widetilde{a^{-1}})SH(a^{-1})$$

are regularizers of $T(a^{-1})$ and $T(\tilde{a})$, respectively. A few application of 1.12(b), 2.7 and 1.4(b) shows that in this case $\operatorname{Ind} T(\tilde{a}) = \operatorname{Ind} T(a^{-1})$. Moreover, the Theorems 2.35 and 2.38 below imply that for $a \in (C + H^\infty)_{N\times N}$ or $a \in PC_{N\times N}$ the equivalence

$$T(a) \in \Phi(H^2_N) \Longleftrightarrow T(\tilde{a}) \in \Phi(H^2_N)$$

holds and that then $\operatorname{Ind} T(a) = -\operatorname{Ind} T(\tilde{a})$. The latter equivalence can easily be shown to be true for general $a \in L^\infty$ in the scalar case, but in the matrix case we do not know whether it is true for general $a \in L^\infty_{N\times N}$ or not.

<u>2.35. Theorem.</u> Let $a \in (C + H^\infty)_{N\times N}$. Then $T(a) \in \Phi(H^2_N)$ if and only if $\det a \in G(C + H^\infty)$. In that case

$$\operatorname{Ind} T(a) = \operatorname{Ind} T(\det a),$$

where $\operatorname{Ind} T(a)$ refers to the index of $T(a)$ on H^2_N and $\operatorname{Ind} T(\det a)$ to that of $T(\det a)$ on H^2.

<u>Proof.</u> By virtue of Proposition 2.7 and Theorem 2.16 we have

$$T(\varphi)T(\psi) - T(\psi)T(\varphi) = H(\psi)H(\tilde{\varphi}) - H(\varphi)H(\tilde{\psi}) \in \mathfrak{C}_\infty(H^2)$$

for all $\varphi, \psi \in C + H^\infty$. Therefore the entries of the operator matrix $T(a) = (T(a_{jk}))_{j,k=1}^N$ commute modulo $\mathfrak{C}_\infty(H^2)$. According to the corollary in 1.15

$$\begin{aligned} T(a) \in \Phi(H^2_N) &\Longleftrightarrow \det(T(a_{jk})) \in \Phi(H^2) \\ &\Longleftrightarrow T(\det a) \in \Phi(H^2), \end{aligned}$$

which by Theorem 2.18 is equivalent to $\det a \in G(C + H^\infty)$.
It remains to prove the index formula.
Let $a_{jk} = f_{jk} + h_{jk}$, $f_{jk} \in C$, $h_{jk} \in H^\infty$, and put $f = (f_{jk})_{j,k=1}^N$, $h = (h_{jk})_{j,k=1}^N$.
Approximating each f_{jk} by a trigonometric polynomial p_{jk} sufficiently close in the L^∞-norm, setting then $p = (p_{jk})_{j,k=1}^N$ and taking into consideration 1.12(d), we get

$$\operatorname{Ind} T(a) = \operatorname{Ind} T(f + h) = \operatorname{Ind} T(p + h)$$

and

$$\operatorname{Ind} T(\det a) = \operatorname{Ind} T(\det(f + h)) = \operatorname{Ind} T(\det(p + h)).$$

Finally, for any trigonometric polynomials p, q and for any $g, h \in H^{\infty}$ there holds

$$T(p + g)T(q + h) - T(q + h)T(p + g)$$
$$= H(q + h)H(\tilde{p}) - H(p + g)H(\tilde{q}) \in \mathfrak{C}_0(H^2) \subset \mathfrak{C}_1(H^2);$$

so the theorem in 1.16 applies and it results

$$\mathrm{Ind}\, T(p + h) = \mathrm{Ind}\, T(\det(p + h)),$$

which completes the proof. Δ

Remark. By Remark 1 in 2.18, we actually have

$$\mathrm{Ind}\, T(a) = \mathrm{Ind}\, T(\det a) = - \mathrm{ind} \det a.$$

2.36. Theorem. Let $a \in (C \cap M_{<p>})_{N\times N}$, $1 \leqq p < \infty$. Then $T(a) \in \Phi(l^p_N)$ if and only if $\det a(t) \neq 0$ for all $t \in \Gamma$. In that case

$$\mathrm{Ind}\ T(a) = - \mathrm{ind} \det a.$$

Proof. Due to Propositions 2.7 and 2.20 the commutator of Toeplitz operators with symbols from $C \cap M_{<p>}$ is compact on l^p. Thus the corollary in 1.15 may be applied to show that $T(a) \in \Phi(l^p_N)$ if and only if $T(\det a) \in \Phi(l^p)$, which on its hand by Theorem 2.21 is equivalent to the conditions of the present theorem.
The index formula follows by an appropriate approximation argument as in the proof of Theorem 2.35 from the theorem in 1.16 (take the Fejer means of the Fourier series, cf. Verbickiĭ, Krupnik [1]). Δ

Remark. The condition that $\det a(t) \neq 0$ for all $t \in \Gamma$ is necessary for $T(a)$ to be a Φ_+- or Φ_--operator (cf. 1.15 and Remark 2 in 2.21).

2.37. Lemma. Suppose $a \in (PC \cap M_{<p>})_{N\times N}$, $1 < p < \infty$, and $\det a(t \pm 0) \neq 0$ for all $t \in \Gamma$. Then the matrix-function $a(t)$ admits a factorization $a = b\varphi c$, where $b, c \in (C \cap M_{<p>})_{N\times N}$ are regular matrix-functions and where $\varphi \in (PC \cap M_{<p>})_{N\times N}$ is an upper triangular matrix-function.

Proof. Let $t_1, \ldots, t_R$ denote the points of discontinuity of $a(t)$. Choose regular constant matrices d_r such that

$$d_r a^{-1}(t_r + 0)a(t_r - 0)d_r^{-1}$$

become upper triangular matrices $(r = 1, \ldots, R)$. Then take $d \in (C \cap M_{<p>})_{N\times N}$ such that $d(t_r) = d_r$ $(r = 1, \ldots, R)$. Finally, let $\varphi \in (PC \cap M_{<p>})_{N\times N}$ be an upper triangular matrix with the following properties: φ is continuous for $t \notin \{t_1,\ldots,t_R\}$, $\det \varphi(t \pm 0) \neq 0$ for all $t \in \Gamma$, $\varphi(t_r) = I_{N\times N}$, and

$$\varphi(t_r + 0) = d_r a^{-1}(t_r - 0)a(t_r + 0)d_r^{-1} \quad (r = 1,\ldots,R).$$

Then

$$a(t_r + 0)d_r^{-1}\varphi^{-1}(t_r + 0) = a(t_r - 0)d_r^{-1}\varphi^{-1}(t_r),$$

hence $c = ad^{-1}\varphi^{-1} \in (C \cap M_{<p>})_{N\times N}$, which completes the proof. Δ

Now, given $a \in (\overline{PC} \cap M_{<p>})_{N\times N}$ define for $(t,\mu) \in \Gamma \times [0,1]$

$$a_p(t,\mu) := (1 - s_p(\mu))a(t - 0) + s_p(\mu)a(t + 0)$$

(see 2.31). Note that $\det[a_p(t,\mu)]$ traces out a continuous, closed, and oriented curve in $\mathbb{C}$ when (t,μ) is running through the cylinder $\Gamma \times [0,1]$.

<u>2.38. Theorem.</u> Let $a \in (PC \cap M_{<p>})_{N\times N}$, $1 < p < \infty$. Then $T(a) \in \Phi(l_N^p)$ if and only if $\det[a_p(t,\mu)] \neq 0$ for all $(t,\mu) \in \Gamma \times [0,1]$. In that case

$$\operatorname{Ind} T(a) = - \operatorname{ind} \det[a_p(t,\mu)]. \tag{20}$$

<u>Proof.</u> Suppose $\det[a_p(t,\mu)] \neq 0$ for all $(t,\mu) \in \Gamma \times [0,1]$. Let $a = b\varphi c$ be the factorization of $a(t)$ in accordance with the preceding lemma. Then $\det[\varphi_p(t,\mu)] \neq 0$ and, consequently, we have $\varphi_p^j(t,\mu) \neq 0$ for the diagonal entries φ^j $(j = 1,\ldots,N)$ of φ.
By the theorem in 2.31 we get $T(\varphi^j) \in \Phi(l^p)$, and therefore $T(\varphi) \in \Phi(l_N^p)$. Finally

$$T(a) - T(b)T(\varphi)T(c) \in \mathfrak{C}_\infty(l_N^p),$$

and since by Theorem 2.36 $T(b) \in \Phi(l_N^p)$ and $T(c) \in \Phi(l_N^p)$, it follows that $T(a) \in \Phi(l_N^p)$.
For the proof of the index formula we only remark that

$$\operatorname{Ind} T(b) = - \operatorname{ind} \det b = - \operatorname{ind} \det b_p(t,\mu),$$
$$\operatorname{Ind} T(c) = - \operatorname{ind} \det c = - \operatorname{ind} \det c_p(t,\mu)$$

(by Theorem 2.36) and that

$$\operatorname{Ind} T(\varphi) = \sum_{j=1}^{N} \operatorname{Ind} T(\varphi^j) = - \sum_{j=1}^{N} \operatorname{ind} \varphi_p^j(t,\mu)$$
$$= - \operatorname{ind} \prod_{j=1}^{N} \varphi_p^j(t,\mu) = - \operatorname{ind} \det[\varphi_p(t,\mu)]$$

(cf. 2.31(16)). Now (20) follows by applying 1.12(c).

It remains to prove the necessity portion of our theorem. If $T(a) \in \Phi(l^p_N)$, then $\det a(t \pm 0) \neq 0$ for all $t \in \Gamma$ by Theorem 2.33. Let $a = b\varphi c$ be the factorization in correspondence with Lemma 2.37. Then

$$T(a) - T(b)T(\varphi)T(c) \in \mathfrak{C}_\infty(l^p_N)$$

and the Fredholmness of $T(a)$, $T(b)$, $T(c)$ implies that of $T(\varphi)$. Since φ is a triangular matrix, we get $T(\varphi^j) \in \Phi(l^p)$ for the diagonal entries φ^j of φ. Thus $\det[\varphi_p(t,\mu)] \neq 0$ for all $(t,\mu) \in \Gamma \times [0,1]$ by the theorem in 2.31. It results that

$$\det[a_p(t,\mu)] = \det b(t) \cdot \det[\varphi_p(t,\mu)] \cdot \det c(t) \neq 0$$

for all $(t,\mu) \in \Gamma \times [0,1]$, and the proof is complete. Δ

Remark 1. This theorem remains valid for $a \in (\overline{PC} \cap M_{<p>})_{N\times N}$. The proof can be given by applying the usual localization techniques.

Remark 2. The conditions of the theorem are also necessary for $T(a)$ to be a Φ_+- or Φ_--operator on l^p_N. This follows by an appropriate perturbation argument.

Remark 3. Since the commutator of Toeplitz operators with piecewise continuous symbols from $M_{<p>}$ is compact, the corollary in 1.15 gives

$$T(a) = (T(a_{jk}))^N_{j,k=1} \in \Phi(l^p_N)$$

$$\Longleftrightarrow \det(T(a_{jk}))^N_{j,k=1} \in \Phi(l^p). \tag{21}$$

In the case of continuous symbols (cf. the proof of Theorem 2.36) we had

$$\det(T(a_{jk}))^N_{j,k=1} - T(\det(a_{jk})^N_{j,k=1}) \in \mathfrak{C}_\infty(l^p)$$

and so (21) was equivalent to $T(\det a) \in \Phi(l^p)$. But the quasicommutator of Toeplitz operators with piecewise continuous symbols is, in general, not compact. Thus (21) suggests to study objects which are sums of products of Toeplitz operators, and therefore leads to the investigation of algebras generated by Toeplitz operators.

ALGEBRAS GENERATED BY TOEPLITZ OPERATORS

2.39. Suppose $\mathcal{A}$ is a Banach algebra, X is a Banach space and $i : \mathcal{A} \to \mathcal{B}(X)$ is a continuous imbedding of $\mathcal{A}$ into $\mathcal{B}(X)$, i.e. i is a continuous linear operator with trivial kernel and closed image.

For $a \in \mathcal{A}$ we shall write $T(a)$ for the element $i(a) \in \mathcal{B}(X)$. Now let us denote by

$T(\mathcal{A})$ the set $i(\mathcal{A})$;

$\operatorname{alg} T(\mathcal{A})$ the closed subalgebra of $\mathcal{B}(X)$ generated by $T(\mathcal{A})$, that is

$$\operatorname{alg} T(\mathcal{A}) = \operatorname*{clos}_{\mathcal{B}(X)}\{ \sum_{i=1}^{n} \prod_{j=1}^{m} T(a_{jk}) : a_{jk} \in \mathcal{A}\} ;$$

$Q(\mathcal{A})$ the quasicommutator ideal in $\operatorname{alg} T(\mathcal{A})$, i.e. the smallest closed two-sided ideal in $\operatorname{alg} T(\mathcal{A})$ containing all elements of the form

$$T(ab) - T(a)T(b), \quad a \in \mathcal{A}, \quad b \in \mathcal{A}.$$

In addition assume there is a constant $\gamma > 0$ such that

$$\|T(\sum_{i=1}^{n} \prod_{j=1}^{m} a_{ij})\|_{\mathcal{B}(X)} \leq \gamma \,\|\sum_{i=1}^{n} \prod_{j=1}^{m} T(a_{ij})\|_{\mathcal{B}(X)} \tag{22}$$

for every $n,m \in \mathbb{Z}_+$ and $a_{ij} \in \mathcal{A}$ holds.

<u>2.40. Theorem.</u> Under the assumption (22) the algebra $\operatorname{alg} T(\mathcal{A})$ decomposes into a direct sum of $T(\mathcal{A})$ and $Q(\mathcal{A})$:

$$\operatorname{alg} T(\mathcal{A}) = T(\mathcal{A}) \oplus Q(\mathcal{A}).$$

<u>Proof.</u> The set

$$\mathfrak{M} = \{\sum_{i=1}^{m} \prod_{j=1}^{n} T(a_{ij}) : n,m \in \mathbb{Z}_+, \quad a_{ij} \in \mathcal{A}\}$$

is dense in $\operatorname{alg} T(\mathcal{A})$. Define the linear mapping S on $\mathfrak{M}$ by

$$S : \sum_{i=1}^{m} \prod_{j=1}^{n} T(a_{ij}) \mapsto T(\sum_{i=1}^{m} \prod_{j=1}^{n} a_{ij}).$$

By (22), S is correctly defined and continuous. It follows from $S(T(a)) = T(a)$ that S is a projection on $\mathfrak{M}$. Thus S admits an extension to a continuous projection on $\operatorname{alg} T(\mathcal{A})$. It is easy to see that $S(\operatorname{alg} T(\mathcal{A})) = T(\mathcal{A})$.

Now we claim that $\operatorname{Ker} S$ is an (obviously, closed) two-sided ideal in $\operatorname{alg} T(\mathcal{A})$.
Let $A \in \operatorname{Ker} S$, $B \in \operatorname{alg} T(\mathcal{A})$. Then there are sequences $\{A_n\}$, $\{B_n\} \subset \mathfrak{M}$ such that (here $\rightrightarrows$ denotes uniform convergence)

$$A_n \rightrightarrows A, \quad B_n \rightrightarrows B, \quad S(A_n) \rightrightarrows 0.$$

Hence $A_nB_n \rightrightarrows AB$, and since by (22)

$$\|S(A_nB_n)\| \leq \gamma\|S(A_n)S(B_n)\| \leq \gamma\|S(A_n)\|\,\|S(B_n)\|,$$

we get $S(AB) = 0$. Analogously, $S(BA) = 0$.

Thus S is a closed two-sided ideal in $\operatorname{alg} T(\mathcal{A})$.

Finally, let us prove that $\operatorname{Ker} S = Q(\mathcal{A})$. The inclusion $Q(\mathcal{A}) \subset \operatorname{Ker} S$ is obvious. Let now $A \in \operatorname{Ker} S$. Again there is a sequence $\{A_n\} \subset \mathfrak{A}$ such that $A_n \rightrightarrows A$ and $S(A_n) \to 0$. A little thought shows that

$$A_n - S(A_n) \in Q(\mathcal{A})$$

and passing to the limit $n \to \infty$ we obtain $A \in Q(\mathcal{A})$. Thus $\operatorname{Ker} S \subset$
$\subset Q(\mathcal{A})$.
From $\operatorname{alg} T(\mathcal{A}) = \operatorname{Ker} S \oplus \operatorname{Im} S$, $\operatorname{Im} S = T(\mathcal{A})$, $\operatorname{Ker} S = Q(\mathcal{A})$, the assertion follows. Δ

Corollary 1. The sequence

$$(0) \to Q(\mathcal{A}) \to \operatorname{alg} T(\mathcal{A}) \to \mathcal{A} \to (0)$$

is short exact; that is, the quotient algebra $\operatorname{alg} T(\mathcal{A})/Q(\mathcal{A})$ is isomorphic to $\mathcal{A}$.

Corollary 2. For $a \in \mathcal{A}$ and with $\varrho(\cdot)$ as notation for the spectral radius we have

$$\varrho(a) \leqq \varrho(T(a)) \leqq \|i\|\|a\|_{\mathcal{A}}.$$

2.41. Applications and examples.

(a) Let $\mathcal{A} = M_p$, $X = l^p$ and let i assign to each $a \in M_p$ the Toeplitz operator $T(a) \in \mathcal{B}(l^p)$. Then

$$\operatorname{alg} T(M_p) = T(M_p) \oplus Q(M_p).$$

Proof. It remains to verify 2.39(22) for the case considered here. This can be done as in the proof of Theorem 2.10:

Taking into consideration that $M(\chi_n) \in \mathcal{B}(\tilde{l}^p)$ is an isometry and that

$$\operatorname*{s-lim}_{n\to\infty} M(\chi_{-n})PM(a)PM(\chi_n) = M(a),$$

we get (the sums and products finite)

$$\|\sum_i \prod_j T(a_{ij})\|_{\mathcal{B}(l^p)} = \|\sum_i \prod_j PM(a_{ij})P\|_{\mathcal{B}(\tilde{l}^p)}$$

$$= \|M(\chi_{-n})(\sum_i \prod_j PM(a_{ij})P)M(\chi_n)\|_{\mathcal{B}(\tilde{l}^p)}$$

$$= \|\sum_i \prod_j (M(\chi_{-n})PM(a_{ij})PM(\chi_n))\|_{\mathcal{B}(\tilde{l}^p)}$$

$$\geq \|\sum_i \prod_j M(a_{ij})\|_{\mathcal{B}(\widetilde{l}^p)} = \|M(\sum_i \prod_j a_{ij})\|_{\mathcal{B}(\widetilde{l}^p)}$$

$$= \|T(\sum_i \prod_j a_{ij})\|_{\mathcal{B}(\widetilde{l}^p)} ,$$

where the last equality results from Theorem 2.5. Since finite sums of finite products of Toeplitz operators are dense in alg $T(M_p)$, we obtain the assertion. Δ

(b) For $n \geqq 0$ put $V^n = T(\chi_n)$, $V^{(-n)} = T(\chi_{-n})$. Then for every $A \in$ alg $T(M_p)$ the strong limit $\text{s-}\lim_{n\to\infty} V^{(-n)}AV^n$ exists and is equal to $S(A)$, where S was defined in the proof of Theorem 2.40. In particular, for $A \in$ alg $T(M_p)$ we have

$$A \in Q(M_p) \iff \text{s-}\lim_{n\to\infty} V^{(-n)}AV^n = 0.$$

Proof. The same argument as in (a) applies:

$$\text{s-}\lim_{n\to\infty} T(\chi_{-n})(\sum_i \prod_j T(a_{ij}))T(\chi_n)$$

$$= \text{s-}\lim_{n\to\infty} PM(\chi_{-n})(\sum_i \prod_j PM(a_{ij})P)M(\chi_n)P$$

$$= PM(\sum_i \prod_j a_{ij})P. \ \Delta$$

(c) Since $V^n \to 0$ weakly, we obtain that $KV^n \to 0$ strongly for every $K \in \mathfrak{C}_\infty(l^p)$. Thus, combining (a) and (b),

$$\text{alg } T(M_p) \cap \mathfrak{C}_\infty(l^p) \subset Q(M_p).$$

In particular, the only compact Toeplitz operator is 0.

(d) A trivial example: let $\mathfrak{A} = H^\infty$, $X = H^2$ and let i be as in (a). Then $Q(H^\infty) = \{0\}$ and

$$\text{alg } T(H^\infty) = T(H^\infty) \cong H^\infty.$$

(e) Let $\mathfrak{A} = C$ or $C + H^\infty$ or $(C + H^\infty) \cap (C + \overline{H^\infty})$, $X = H^2$ and let i be as in (a). Then $Q(\mathfrak{A}) = \mathfrak{C}_\infty(H^2)$, consequently,

$$\text{alg } T(C) = T(C) \oplus \mathfrak{C}_\infty(H^2),$$
$$\text{alg } T(C + H^\infty) = T(C + H^\infty) \oplus \mathfrak{C}_\infty(H^2)$$
$$\text{alg } T((C + H^\infty) \cap (C + \overline{H^\infty}))$$
$$= T((C + H^\infty) \cap (C + \overline{H^\infty})) \oplus \mathfrak{C}_\infty(H^2).$$

Proof. Due to (a) the estimate 2.39(22) holds. Since $T(ab)-T(a)T(b)= = H(a)H(\tilde{b})$ is compact for $a,b \in C + H^\infty$ (see Theorem 2.16), we get

$$Q(C) \subset Q((C + H^\infty) \cap (C + \overline{H^\infty})) \subset Q(C + H^\infty) \subset \mathfrak{C}_\infty(H^2).$$

It remains to show that $\mathfrak{C}_\infty(H^2) \subset Q(C)$. For this it suffices to prove that every finite-rank operator belongs to $Q(C)$. This, on its part, will follow, once we have shown that every operator K of the form $K\varphi = (\varphi,p)q$, where p and q are trigonometric polynomials in H^∞, belongs to $Q(C)$.
But this is obvious in view of the identity

$$K = T(q)[T(\chi_1\chi_{-1}) - T(\chi_1)T(\chi_{-1})]T(\overline{p}). \ \Delta$$

(f) Since $\operatorname{alg} T(C)$ and $\operatorname{alg} T((C + H^\infty) \cap (C + \overline{H^\infty}))$ are C*-algebras, the argument of (e) gives that

$$T(a) \in \Phi(H^2) \Longleftrightarrow a \in GC$$

for $a \in C$, and that

$$T(a) \in \Phi(H^2) \Longleftrightarrow a \in G((C + H^\infty) \cap (C \cap \overline{H^\infty}))$$

for $a \in (C + H^\infty) \cap (C + \overline{H^\infty})$.

2.42. Some remarks about piecewise continuous symbols.

Let $\operatorname{alg} T(PC)$ denote the Banach algebra generated in $\mathcal{B}(H^2)$ by Toeplitz operators with symbols from PC. Obviously, $\operatorname{alg} T(PC) = \operatorname{alg} T(\overline{PC})$.
Proposition 2.25 and the proof of 2.41(e) imply that the commutator ideal of $\operatorname{alg} T(\overline{PC})$ is just $\mathfrak{C}_\infty(H^2)$. Concerning the quasicommutator ideal $Q(PC) = Q(\overline{PC})$ we have, again by 2.41(e), $\mathfrak{C}_\infty(H^2) \subset Q(\overline{PC})$, but we emphasize that there are non-compact operators in $Q(\overline{PC})$. Nevertheless, as will be outlined below, the quotient algebra $\operatorname{alg} T(\overline{PC})/\mathfrak{C}_\infty(H^2)$ admits a very nice description.

Let us turn over to the l^p-case. Denote by $\overline{PC}_p$ $(1 < p < \infty)$ the closure of the algebra of all piecewise constant functions on the unit circle with respect to the norm $\|a\| = \|M(a)\|_{\mathcal{B}(\hat{l}^p)}$. Then put

$$\operatorname{alg} T(\overline{PC}_p) = \operatorname*{clos}_{\mathcal{B}(l^p)} \{ \sum_{i=1}^{n} \prod_{j=1}^{m} T(a_{ij}) : n,m \in \mathbb{Z}_+,\ a_{ij} \in PC_p \}.$$

It can be shown that $\mathfrak{C}_\infty(l^p)$ forms a closed two-sided ideal in $\operatorname{alg} T(\overline{PC}_p)$, that $\operatorname{alg} T(\overline{PC}_p)/\mathfrak{C}_\infty(l^p)$ is commutative, and that its maximal ideal space is homeomorphic to the cylinder $\mathfrak{M} = \Gamma \times [0,1]$ equipped with an exotic topology. For $a \in \overline{PC}_p$ the Gelfand transformation of the coset in the quotient algebra containing $T(a)$ is just

$$(1 - s_p(\mu))a(t - 0) + s_p(\mu)a(t + 0),\ (t,\mu) \in \mathfrak{M}$$

(recall the notation in 2.31 and write $s_2(\mu) = \mu$ for $p = 2$).

In the case $p = 2$ we have $\overline{PC}_2 = \overline{PC}$ and since $\operatorname{alg} T(\overline{PC})/\mathbb{C}_\infty(H^2)$ is a C^*-algebra, the Gelfand-Naimark theorem applies. What results is the following.

Theorem (Gohberg-Krupnik). The quotient algebra $\operatorname{alg} T(\overline{PC})/\mathbb{C}_\infty(H^2)$ is isometrically isomorphic to $C(\mathfrak{A})$.

NOTES AND COMMENTS

The best general references are Douglas [2], [3], [4], Nikol'skiǐ [3], Gohberg, Feldman [1], Gohberg, Krupnik [2], Prößdorf [1].

2.2. See Verbickiǐ [1] for the present status.

2.3. To the general references above the survey Power [2] should be added, where one can also find historical remarks.

2.5. - 2.6. For $p = 2$ see Brown, Halmos [1] and Halmos [1]; in the case $p \neq 2$ these results essentially go back to Dudučava [3]. See also Dudučava [6] and Gohberg, Krupnik [2]. Obviously, Theorem 2.5 also holds for $p = 1$.

2.8. This theorem was found by Coburn [1] for $p = 2$, and proved in Dudučava [3] for $p \neq 2$.

2.10. - 2.11. See Hartman, Wintner [1] and Brown, Halmos [1] for $p=2$.

2.12. In the H^∞-case see Wintner [1] for the spectrum (actually we prove this result in 2.41(d) and 4.7) and Nikol'skiǐ [3] for the essential spectrum. The spectrum of self-adjoint Toeplitz operators was identified in Hartman, Wintner [1]. The assertion on the essential spectrum follows from its connectedness. Note that for arbitrary $a \in L^\infty$ the spectra $\sigma(T(a))$ (Widom [4]) and $\sigma_{ess}(T(a))$ (Douglas [2], 7.45) are always connected.

2.13. See Nehari [1]; for recent works we refer to Nikol'skiǐ [2], VIII.

2.14. - 2.15. These results are essentially due to Widom [2] and Devinatz [1], but see also Helson, Szegö [1]. For the matrix case see Rabindranathan [1] and Douglas [3].

2.16. The compactness criterion goes back to Hartman [1]; the essential norm of a Hankel operator was first determined by Adamian, Arov and Krein.

2.17. (a) Kronecker [1], see also Peller, Hruščev [1] and Nikol'skiǐ [2], VIII; (b) Peller [1], [2], [3], confer also Peller, Hruščev [1], [2]; (c) Adamian, Arov, Krein [1].

<u>2.18.</u> See Douglas [1]. We refer to Douglas [2], [3], [4] and especially to Nikol'skiĭ [3] for a further discussion and for related topics. The proof of the text is taken from Peller, Hruščev [1]. Another proof is given in 4.5.

<u>2.20.</u> This proof is adopted from Gohberg, Krupnik [2].

<u>2.21.</u> This theorem is the culmination of several authors including Michlin [1], Gohberg [1], Simonenko [1] (in the language of singular integral operators), Krein [1], Calderon, Spitzer, Widom [1], Widom [1] and Devinatz [1]. The proof presented in the text is patterned after Gohberg, Krupnik [2].

<u>2.23. - 2.26.</u> The Propositions 2.24 and 2.25 are well-known; see e.g. Dudučava [3], [6] and Gohberg, Krupnik [2]. In Basor [2] and Böttcher [1] the following result was proved: if $a,b \in L^\infty$ and if $f + g = 1$ is a smooth partition of unity such that af and bg both are in $C^{1+\varepsilon}$ $(\varepsilon > 0)$, then $T(ab) - T(a)T(b) \in \mathfrak{C}_1(l^2)$. Proposition 2.26 and its corollaries are a stronger version of this result and are published here for the first time. The method used to derive the Proposition 2.24 and 2.26 was suggested by Basor [2].

<u>2.27.</u> (a) Brown, Halmos [1], Halmos [1] (Problem 195); (b) "<==" was proved in Axler, Chang, Sarason [1], "==>" in Vol'berg [1], see also Nikol'skiĭ [3], some special cases (including Proposition 2.24) were already considered in Gohberg, Krupnik [1], [2]; (c) note that actually

$$\operatorname{rank} H(a)H(\tilde{b}) = \min\{\operatorname{rank} H(a), \operatorname{rank} H(\tilde{b})\},$$

see Axler, Chang, Sarason [1], but also Nikol'skiĭ [3], an "algebraical" proof was recently found by Richmann [1].

<u>2.28. - 2.31.</u> For the case $p = 2$ these results go back to Widom [1] and Devinatz [1], but are also contained in the more general results of Simonenko [1]; see also Gohberg [2]. The results for $p \neq 2$ are due to Dudučava [2], [3]; see also Gohberg, Krupnik [2].

<u>2.33.</u> Note that we require only $a \in M_p$, and not $a \in M_{<p>}$.

<u>2.35.</u> See Douglas [3], Theorem 6. Confer also the remark in 4.5.

<u>2.36.</u> Here should be referred to Gohberg, Krein [1].

<u>2.37.</u> This lemma was first stated in Gohberg, Krupnik [1].

<u>2.38.</u> This theorem is essentially due to Dudučava, but see also Simonenko [2].

2.39. - 2.41. Some of the results of these sections are published here for the first time. The reader should also confer Roch, Silbermann [2]. For a detailed study of algebras generated by Toeplitz operators (at least for $p = 2$) we refer to Douglas [2], [3], [4], Nikol'skiĭ [3], Gohberg, Krupnik [1].
Under somewhat more restrictive hypotheses, the Corollary 1 of 2.40 in this abstract version was stated in Clancey [1]. We draw the reader's attention to the role played by the "shifts" $U_{\pm}^{n}$ and $V^{(\pm n)}$ (cf. 2.5, 2.10, 2.16, 2.33, 2.41, but also 3.9 below). Brown, Halmos [1] and Nikol'skiĭ [3] have come very close to using them in this context, but, as far as we know, their actual significance was thoroughly pointed out in Clancey [1] for the first time. However, the approach proposed in the text seems to be more natural than that (already very simple) of Clancey [1]. Note that the limit $\text{s-lim}_{n\to\infty} V^{(-n)}AV^{n}$ plays an important part in the theory of the finite section method, too (see 3.9 and confer Roch, Silbermann [2] for a further discussion).
The elegant proof of the inclusion $\mathfrak{C}_{\infty}(H^2) \subset Q(C)$ presented in 2.41(e) is due to Coburn [2].
In connection with 2.41(f) see also 4.12. The algebra $(C+H^{\infty}) \cap (C+\overline{H^{\infty}})$ is called the algebra of quasicontinuous functions and denoted by QC. The algebra $PQC := \text{alg}(PC, QC)$ is usually referred to as the algebra of piecewise quasicontinuous functions. Toeplitz and Hankel operators with PQC-symbols were studied in Sarason [2] and Power [1].
Finally, once more we recommend the reader to confer the excellent article Douglas [4] for a host of further sharp results along these lines, obtained on the basis of a very sensitive localization technique.

2.42. See Dudučava [3], [4], and Gohberg, Krupnik [1].

CHAPTER 3

THE FINITE SECTION METHOD

PROJECTION METHODS FOR THE SOLUTION OF OPERATOR EQUATIONS

3.1. Projection methods. Let X and X' be Banach spaces and let $A \in \mathcal{B}(X,X')$. One method for the approximate solution of the equation

$$Ax = y \tag{1}$$

can be described as follows. Let $\{P_n\}$ and $\{P'_n\}$ be sequences of projections $P_n \in \mathcal{B}(X)$ and $P'_n \in \mathcal{B}(X')$ such that $P_n \to I_X$ and $P'_n \to I_{X'}$ strongly, and let $A_n : \operatorname{Im} P_n \to \operatorname{Im} P'_n$ be certain given bounded operators. Frequently we think of A_n as identified with A_nP_n and may therefore regard A_n as an element of $B(X,X')$. Now consider the equation

$$A_nx_n = P'_ny, \quad x_n \in \operatorname{Im} P_n. \tag{2}$$

We write $A \in \Pi_{X,X'}\{A_n\}$ if for every $y \in X'$, there is an n_0 (depending, in general, on y) such that equation (2) has a unique solution $x_n \in \operatorname{Im} P_n$ for all $n \geqq n_0$ and if x_n converges in the norm of X to a solution $x \in X$ of the equation (1). If $X = X'$ and $P_n = P'_n$ then $\Pi_{X,X}\{A_n\}$ will be abbreviated to $\Pi_X\{A_n\}$, or even to $\Pi\{A_n\}$ if this does not lead to misunderstandings.

3.2. Notations. We denote by

$\mathcal{F}$ the collection of all sequences $\{A_n\}_{n=0}^{\infty}$ of operators $A_n : \operatorname{Im} P_n \to \operatorname{Im} P_n$ such that $\sup_n \|A_nP_n\|_{\mathcal{B}(X)} < \infty$; defining $\{A_n\} + \{B_n\} = \{A_n + B_n\}$, $\{A_n\}\{B_n\} = \{A_nB_n\}$ and $\|\{A_n\}\|_{\mathcal{F}} = \sup_n \|A_nP_n\|_{\mathcal{B}(X)}$ the set $\mathcal{F}$ becomes a Banach algebra;

$\mathcal{F}_c$ the closed subalgebra of $\mathcal{F}$ consisting of all sequences $\{A_n\}$ which possess a strong limit $A \in \mathcal{B}(X)$;

$\mathcal{N}$ the collection of all sequences $\{C_n\} \in \mathcal{F}$ such that $\|C_nP_n\|_{\mathcal{B}(X)} \to 0$ as $n \to \infty$; it is easy to see that $\mathcal{N}$ forms a closed two-sided ideal in both $\mathcal{F}$ and $\mathcal{F}_c$;

π the canonical projection of $\mathcal{F}$ onto $\mathcal{F}/\mathcal{N}$;

π_c the canonical projection of $\mathcal{F}_c$ onto $\mathcal{F}_c/\mathcal{N}$.

In what follows $\to$ denotes strong convergence and $\rightrightarrows$ uniform convergence (operator-norm convergence); $o(1)$ denotes any sequence in $\mathcal{N}$.

3.3. Proposition. Suppose $P_n \to I$ and $A_nP_n \to A$ as $n \to \infty$. Then the following statements are equivalent:

(a) $A \in \Pi\{A_n\}$.

(b) $A \in G\mathcal{B}(X)$, the operators $A_n : \operatorname{Im} P_n \to \operatorname{Im} P_n$ are invertible for all sufficiently large n (say $n \geqq n_0$) and

$$\sup_{n \geqq n_0} \|A_n^{-1}P_n\|_{\mathcal{B}(X)} < \infty.$$

(c) $\pi_c\{A_n\} \in G(\mathcal{F}_c/\mathcal{N})$.

(d) $A \in G\mathcal{B}(X)$ and $\pi\{A_n\} \in G(\mathcal{F}/\mathcal{N})$.

Proof. (a) ==> (b) The Banach-Steinhaus theorem gives that $\sup_n \|A_n^{-1}P_n\| =: C < \infty$ and therefore

$$\|P_nx\| \leqq C\|A_nP_nx\| \tag{3}$$

for all $x \in X$. Passage to the limit $n \to \infty$ in (3) leads to $\|x\| \leqq C\|Ax\|$, that is, $\operatorname{Ker} A = \{0\}$. The definition of $\Pi\{A_n\}$ involves that $\operatorname{Im} A = X$. Thus $A \in G\mathcal{B}(X)$.

(b) ==> (c) Let $A_n : \operatorname{Im} P_n \to \operatorname{Im} P_n$ be invertible for all $n \geqq n_0$. Put $B_n = P_n$ for $n < n_0$ and $B_n = A_n^{-1}P_n$ for $n \geqq n_0$. Again (3) holds and therefore, for $x \in X$ and $n \geqq n_0$,

$$\|B_nx - P_nA_x^{-1}\| \leqq C\|A_n(B_nx - P_nA_x^{-1})\|$$

$$= C\|P_nx - A_nP_nA_x^{-1}\|.$$

Since $P_n \to I$ and $A_nP_n \to A$, it follows that $B_n \to A^{-1}$. Hence $\{B_n\} \in \mathcal{F}_c$. The assertion is now a consequence of

$$\{B_n\}\{A_n\} - \{P_n\} \in \mathcal{N}, \quad \{A_n\}\{B_n\} - \{P_n\} \in \mathcal{N}.$$

(c) ==> (d) obvious

(d) ==> (a) Suppose

$$\{B_n\} \in \mathcal{F}, \quad B_nA_n = P_n + C_n, \quad A_nB_n = P_n + D_n,$$

$$\{C_n\} \in \mathcal{N}, \quad \{D_n\} \in \mathcal{N}.$$

Then there is an n_0 such that $\|C_n\| < 1/2$ and $\|D_n\| < 1/2$ for all $n \geqq n_0$. Thus $P_n + C_n$ and $P_n + D_n$ regarded as acting on $\operatorname{Im} P_n$ are invertible for $n \geqq n_0$, and the norms of their inverses are uniformly bounded by $1/(1 - 1/2) = 2$. This implies the invertibility of $A_n : \operatorname{Im} P_n \to \operatorname{Im} P_n$ for $n \geqq n_0$ and

$$\sup_{n \geqq n_0} \|A_n^{-1}P_n\| \leqq 2 \sup_{n \geqq n_0} \|B_n\| =: C < \infty.$$

Thus (3) is fulfilled and if

$$Ax = y, \quad A_n x_n = P_n y, \quad x_n \in \operatorname{Im} P_n, \quad n \geqq n_o,$$

then

$$\|x_n - P_n x\| \leqq C\|A_n(x_n - P_n x)\| = C\|P_n y - A_n P_n x\|,$$

which by virtue of $P_n x \to x$, $P_n y \to y$ and $A_n P_n x \to Ax = y$ results in

$$\|x_n - x\| \leqq \|x_n - P_n x\| + \|P_n x - x\| \to 0 \quad (n \to \infty). \; \Delta$$

THE FINITE SECTION METHOD FOR TOEPLITZ OPERATORS

<u>3.4. The finite section method.</u> In the general context of 3.1 put $X = X' = l^p_N$ $(1 \leqq p < \infty)$. Suppose $a \in (M_p)_{N\times N}$ and let $T(a) \in \mathcal{B}(l^p_N)$ be the block Toeplitz operator with the symbol $a(t)$ (recall 2.2 and 2.32). Define the projections P_n $(n = 0,1,2,\ldots)$ on l^p_N by

$$P_n : \{\varphi_0, \varphi_1, \varphi_2, \ldots\} \mapsto \{\varphi_0, \varphi_1, \ldots, \varphi_n, 0, 0, \ldots\}$$

$(\varphi_j \in \mathbb{C}_N)$. Obviously, $P_n \to I$ and $\|P_n\| = 1$.
The operator $P_n T(a) P_n : \operatorname{Im} P_n \to \operatorname{Im} P_n$ may be identified with the finite block Toeplitz matrix $T_n(a) = (a_{j-k})^n_{j,k=o}$, where a_j are the matrix Fourier-coefficients of $a(t)$. Now the projection method 3.1 for the solution of the equation $T(a)\varphi = f$ consists in the transition to the algebraic system

$$T_n(a)\varphi_n = P_n f, \quad \varphi_n \in \operatorname{Im} P_n.$$

Instead of $T(a) \in \Pi\{T_n(a)\}$ we shall write $T(a) \in \Pi\{P_n\}$ or, in order to emphasize the underlying space l^p_N, we use the notation $T(a) \in \Pi_p\{P_n\}$. If $T(a) \in \Pi_p\{P_n\}$, then the <u>finite section method</u> is said to be <u>applicable</u> to $T(a)$ in l^p_N.

<u>3.5. Notations.</u> The operators W_n $(n = 0,1,2,\ldots)$, which are defined on l^p_N by the rule

$$W_n : \{\varphi_0, \varphi_1, \varphi_2, \ldots\} \mapsto \{\varphi_n, \varphi_{n-1}, \ldots, \varphi_0, 0, 0, \ldots\}$$

$(\varphi_j \in \mathbb{C}_N)$ will be of great importance in the following. Obviously,

$$W_n^2 = P_n, \quad W_n P_n = P_n W_n = W_n, \|W_n\| = 1,$$

and W_n converges weakly to 0. Note the identity

$$W_n T_n(a) W_n = T_n(\tilde{a}),$$

where $\tilde{a}$, as usually, is defined by $\tilde{a}(t) = a(1/t)$ $(|t| = 1)$.

Finally put $Q_n = I - P_n$ on l^p_N, that is,

$$Q_n : \{\varphi_0, \varphi_1, \varphi_2, \dots\} \mapsto \{0, \dots, 0, \varphi_{n+1}, \varphi_{n+2}, \dots\}.$$

3.6. Proposition. For $a, b \in (M_p)_{N\times N}$ the following identities hold:

$$P_n T(a) Q_n T(b) P_n = W_n H(\tilde{a}) H(b) W_n, \tag{4}$$

$$T_n(ab) = T_n(a) T_n(b) + P_n H(a) H(\tilde{b}) P_n + W_n H(\tilde{a}) H(b) W_n. \tag{5}$$

Proof. An easy computation shows that

$$W_n T(a) Q_n = P_n H(\tilde{a}) T(\chi_{-n-1}), \tag{6}$$

$$Q_n T(b) W_n = T(\chi_{n+1}) H(b) P_n, \tag{7}$$

and now (4) follows from

$$P_n T(a) Q_n T(b) P_n = W_n W_n T(a) Q_n Q_n T(b) W_n W_n.$$

By Proposition 2.7 we have

$$\begin{aligned} T_n(ab) &= P_n T(ab) P_n \\ &= P_n T(a) T(b) P_n + P_n H(a) H(\tilde{b}) P_n \\ &= P_n T(a) P_n T(b) P_n + P_n T(a) Q_n T(b) P_n \\ &\quad + P_n H(a) H(\tilde{b}) P_n \end{aligned}$$

and (5) results from (4). Δ

3.7. Theorem. Let $a \in (C + H^\infty)_{N\times N}$ and $K \in \mathfrak{C}_\infty(l^2_N)$. Then $T(a) + K \in \Pi_2\{P_n\}$ if and only if $T(a) + K \in G\mathfrak{B}(l^2_N)$ and $T(\tilde{a}) \in G\mathfrak{B}(l^2_N)$. In that case define R_n as

$$R_n = P_n (T(a) + K)^{-1} P_n + W_n (T^{-1}(\tilde{a}) - T(\tilde{a}^{-1})) W_n. \tag{8}$$

Then

$$R_n P_n (T(a) + K) P_n = P_n (T(a) + K) P_n R_n = P_n + o(1) \quad \text{as} \quad n \to \infty,$$

that is, $\pi_c\{R_n\}$ is the inverse of $\pi_c\{P_n(T(a) + K)P_n\}$ in $\mathfrak{F}_c/\mathfrak{N}$.

We shall give three different proofs for this theorem. Here the first, which is direct and needs almost no theory.

Proof. Suppose $T(a) + K \in \Pi_2\{P_n\}$. Then Proposition 3.3 implies that $T(a) + K \in G\mathfrak{B}(l^2_N)$ and

$$\sup_n \| W_n [P_n (T(a) + K) P_n]^{-1} W_n \|$$

$$= \sup_n \| [P_n(T(a) + K)P_n]^{-1} P_n \| =: C < \infty.$$

Consequently,

$$\|P_n\varphi\| \leq C\| W_nP_n(T(a) + K)\ P_nW_nP_n\varphi\| \tag{9}$$

for all $\varphi \in l^2_N$. Since W_n converges weakly to 0 as $n \to \infty$, we get $W_nKW_n \to 0$ (strongly) and passing in (9) to the limit $n \to \infty$ we arrive at the estimate $\|\varphi\| \leq C\|T(\tilde{a})\varphi\|$, holding for all $\varphi \in l^2_N$. Hence $\operatorname{Ker} T(\tilde{a}) = \{0\}$. Now, if $T(a) + K$ is invertible then, as a consequence of Theorem 2.35, $T(\tilde{a})$ is Fredholm and has index zero. Thus, since $T(\tilde{a})$ has a trivial kernel, it must be invertible.
Now suppose $T(a) + K \in G\mathcal{B}(l^2_N)$ and $T(\tilde{a}) \in G\mathcal{B}(l^2_N)$. Then according to Theorem 2.35, $a \in G(C + H^\infty)_{N\times N}$ and $\tilde{a} \in G(C + H^\infty)_{N\times N}$. Defining R_n by (8) we have

$$\begin{aligned}
&R_nP_n(T(a) + K)P_n \\
&= P_n(T(a) + K)^{-1}P_n(T(a) + K)P_n + W_nT^{-1}(\tilde{a})P_nT(\tilde{a})W_n \\
&\quad - P_nT(a^{-1})P_nT(a)P_n + W_n(T^{-1}(\tilde{a})-T(\widetilde{a^{-1}}))W_nKP_n \\
&= P_n - P_n(T(a) + K)^{-1}Q_n(T(a) + K)P_n \qquad (10)\\
&\quad + P_n - W_nT^{-1}(\tilde{a})Q_nT(\tilde{a})W_n \\
&\quad - P_nT(a^{-1})P_nT(a)P_n + W_n(T^{-1}(\tilde{a})-T(\widetilde{a^{-1}}))W_nKP_n.
\end{aligned}$$

Proposition 2.7 and Theorem 2.16 give

$$L := T^{-1}(\tilde{a}) - T(\widetilde{a^{-1}}) = T^{-1}(\tilde{a})H(\tilde{a})H(a^{-1}) \in \mathfrak{C}_\infty(l^2_N) \tag{11}$$

and

$$\begin{aligned}
M &:= (T(a) + K)^{-1} - T(a^{-1}) \\
&= (T(a) + K)^{-1}[H(a)H(\widetilde{a^{-1}}) - KT(a^{-1})] \in \mathfrak{C}_\infty(l^2_N). \qquad (12)
\end{aligned}$$

Hence $W_nLW_n \to 0$ and therefore the last item in (10) is $o(1)$ (cf. 1.3(d)). For the second we obtain in the same fashion

$$\begin{aligned}
&P_n(T(a) + K)^{-1}Q_n(T(a) + K)P_n \\
&= P_nT(a^{-1})Q_nT(a)P_n + o(1) \qquad \text{(by (12))} \\
&= W_nH(\widetilde{a^{-1}})H(a)W_n + o(1) \qquad \text{(by 3.6(4))}.
\end{aligned}$$

Equally, the fourth item becomes

$$W_nT^{-1}(\tilde{a})Q_nT(\tilde{a})W_n$$

$$= W_nT(\widetilde{a^{-1}})Q_nT(\tilde{a})W_n + o(1) \qquad \text{(by (11))}$$

$$= W_nW_nH(a^{-1})H(\tilde{a})W_nW_n + o(1) \qquad \text{(by 3.6.(4))}$$

$$= P_nH(a^{-1})H(\tilde{a})P_n + o(1).$$

Putting these things together and taking into consideration 3.6(5) we get $R_nP_n(T(a) + K)P_n = P_n + o(1)$ as $n \to \infty$.

Analogously can be shown that $P_n(T(a) + K)P_nR_n = P_n + o(1)$ as $n \to \infty$.

Finally, $\{R_n\} \in \mathcal{F}_c$ is obvious from (11). Now Proposition 3.3 completes the proof. Δ

Remark. The computation in the proof of Theorem 3.7 yields the following identity:

$$[P_nT^{-1}(a)P_n + W_n(T^{-1}(\tilde{a}) - T(\widetilde{a^{-1}}))W_n] \cdot T_n(a) =$$

$$= P_n - P_n(T^{-1}(a) - T(a^{-1}))Q_nT(a)P_n \qquad (13)$$

$$- W_n(T^{-1}(\tilde{a}) - T(\widetilde{a^{-1}}))Q_nT(\tilde{a})W_n.$$

This holds, of course, for every $a \in L^\infty_{N\times N}$ such that $T(a), T(\tilde{a}) \in \mathcal{GB}(l^2_N)$ (which implies $a, \tilde{a} \in GL^\infty_{N\times N}$ by Theorem 2.33).

CONSTRUCTIVE THEORY OF THE FINITE SECTION METHOD

3.8. Proposition (Kozak). Let X_1 and X_2 be linear spaces, $A : X_1 \to X_2$ a linear and invertible operator, $P_1 : X_1 \to X_1$ and $P_2 : X_2 \to X_2$ linear projections and put $Q_1 = I - P_1$, $Q_2 = I - P_2$. Then $P_2AP_1 : \operatorname{Im} P_1 \to \operatorname{Im} P_2$ is invertible if and only if $Q_1A^{-1}Q_2 : \operatorname{Im} Q_2 \to \operatorname{Im} Q_1$ is. In that case

$$P_1(P_2AP_1)^{-1}P_2 = P_1A^{-1}P_2 - P_1A^{-1}Q_2(Q_1A^{-1}Q_2)^{-1}Q_1A^{-1}P_2.$$

Proof.

$$P_2AP_1 \cdot (P_1A^{-1}P_2 - P_1A^{-1}Q_2(Q_1A^{-1}Q_2)^{-1}Q_1A^{-1}P_2)$$

$$= P_2AP_1A^{-1}P_2 - P_2AP_1A^{-1}Q_2(Q_1A^{-1}Q_2)^{-1}Q_1A^{-1}P_2$$

$$= P_2AP_1A^{-1}P_2 - P_2AA^{-1}Q_2(Q_1A^{-1}Q_2)^{-1}Q_1A^{-1}P_2$$
$$+ P_2AQ_1A^{-1}Q_2(Q_1A^{-1}Q_2)^{-1}Q_1A^{-1}P_2$$
$$= P_2AP_1A^{-1}P_2 + P_2AQ_1A^{-1}P_2 = P_2. \quad \Delta$$

Corollary. Let $a \in (M_p)_{N\times N}$ and $T(a) \in G\mathcal{B}(l^p_N)$. Then $T(a) \in \Pi_p\{P_n\}$ if and only if $Q_n T^{-1}(a) Q_n : \operatorname{Im} Q_n \to \operatorname{Im} Q_n$ is invertible for all n large enough (say $n \geqq n_o$) and if

$$\sup_{n \geqq n_o} \| (Q_n T^{-1}(a) Q_n)^{-1} Q_n \| < \infty .$$

In that case

$$T_n^{-1}(a) = P_n T^{-1}(a) P_n - P_n T^{-1}(a) Q_n (Q_n T^{-1}(a) Q_n)^{-1} Q_n T^{-1}(a) P_n. \quad (14)$$

3.9. Theorem. Let $A \in \operatorname{alg} T(L^\infty_{N\times N})$ and suppose $A \in \Pi_2\{P_n\}$. Then $S(A^{-1})$ is invertible in $\operatorname{alg} T(L^\infty_{N\times N})$.

Comments. Under our hypotheses, $A \in G\mathcal{B}(l^2_N)$ due to Proposition 3.3. Since $\operatorname{alg} T(L^\infty_{N\times N})$ is a C*-algebra, we conclude that A^{-1} is even in $\operatorname{alg} T(L^\infty_{N\times N})$. Thus

$$S(A^{-1}) = \operatorname*{s-lim}_{n\to\infty} V^{(-n)} A^{-1} V^n,$$

where

$$V^n = T(\chi_n I_{N\times N}) \quad \text{and} \quad V^{(-n)} = T(\chi_{-n} I_{N\times N}),$$

exists and belongs to $\operatorname{alg} T(L^\infty_{N\times N})$ (cf. 2.41(b)).

Proof. Proposition 3.8 implies the invertibility of $Q_n A^{-1} Q_n : \operatorname{Im} Q_n \to \operatorname{Im} Q_n$ for all sufficiently large $n \geqq n_o$ and the uniform boundedness of the norms of the inverses. Thus there is a $\delta > 0$ such that

$$\| Q_n A^{-1} Q_n \varphi \| \geqq \delta \| Q_n \varphi \| \quad (n \geqq n_o)$$

for all $\varphi \in \operatorname{Im} Q_n$. Since $Q_{n-1} = V^n \cdot V^{(-n)}$ and since V^n is an isometry, we obtain

$$\| V^{(-n)} A^{-1} V^n \varphi \| \geqq \delta \| \varphi \| \quad (n \geqq n_o + 1)$$

for all $\varphi \in V^{(-n)} \operatorname{Im} Q_{n-1} = l^2_N$. Passage to the limit $n \to \infty$ gives

$$\| S(A^{-1}) \varphi \| \geqq \delta \| \varphi \|$$

for all $\varphi \in l^2_N$. Next, from 3.3(b) it is easy to deduce that $A^* \in \Pi_2\{P_n\}$ if $A \in \Pi_2\{P_n\}$. The same arguments then lead to

$$\| S((A^*)^{-1}) \varphi \| = \| (S(A^{-1}))^* \varphi \| \geqq \delta \| \varphi \|$$

for all $\varphi \in l^2_N$. Hence $S(A^{-1})$ is invertible in $\mathcal{B}(l^2_N)$ and since $\operatorname{alg} T(L^\infty_{N\times N})$ is a C*-algebra, we obtain $S(A^{-1}) \in G \operatorname{alg} T(L^\infty_{N\times N})$. Δ

Remark 1. Thus as a necessary condition for the applicability of the finite section method to an operator $A \in \text{alg } T(L^{\infty}_{N\times N})$ we obtain besides the invertibility of A itself also the invertibility of the operator $S(A^{-1})$ constructed in the following way: decompose A^{-1} into a sum of an operator in $T(L^{\infty}_{N\times N})$ and one in $Q(L^{\infty}_{N\times N})$, and then annulate the portion belonging to the quasicommutator ideal $Q(L^{\infty}_{N\times N})$. The appearance of the quasicommutator ideal in this context is a somewhat surprising fact. But the role of this ideal in the theory of the finite section method is actually greater than one could expect at first glance.

Remark 2. In Roch, Silbermann [2], it is shown how $S(A)$ can be computed for a sufficiently large class of operators A. We need such a result only in chapter 7. With this single exception, for our further purposes it suffices to know what $S(A)$ looks like if A is of the form $T(a) + K$ or $(T(a) + K)^{-1}$ where K is compact. Obviously,

$$S(T(a)) = \operatorname*{s-lim}_{n\to\infty} V^{(-n)}T(a)V^{n} = T(a),$$

$$S(K) = \operatorname*{s-lim}_{n\to\infty} V^{(-n)}KV^{n} = 0 \quad (K \in \mathfrak{C}_{\infty}(l^{2}_{N}))$$

(note that $V^{n} \to 0$ weakly). Next, the following identity is readily verified:

$$(T(a) + K)^{-1} = T(a^{-1}) + (T(a)+K)^{-1}(T(aa^{-1})-T(a)T(a^{-1})) - (T(a)+K)^{-1}KT(a^{-1}).$$

Thus if $T(a) + K$ is invertible in $\text{alg } T(L^{\infty}_{N\times N})$ and $K \in \mathfrak{C}_{\infty}(l^{2}_{N})$, then by virtue of the inclusion $\mathfrak{C}_{\infty}(l^{2}_{N}) \subset Q(L^{\infty}_{N\times N})$ (cf. 2.41(e))

$$(T(a) + K)^{-1} - T(a^{-1}) \in Q(L^{\infty}_{N\times N})$$

and as a consequence of this (see 2.40)

$$S((T(a) + K)^{-1}) = S(T(a^{-1})) = T(a^{-1}). \tag{15}$$

Corollary. If $a \in L^{\infty}_{N\times N}$, $K \in \mathfrak{C}_{\infty}(l^{2}_{N})$ and $T(a) + K \in \Pi_2\{P_n\}$, then $T(a) + K$ and $T(\tilde{a})$ are invertible on L^{2}_{N}.

Proof. From (15) and the theorem above we get the invertibility of $T(a^{-1})$, which due to Proposition 2.34 is equivalent to that of $T(\tilde{a})$. Δ

Remark 3. We conjecture that the invertibility of both A and $S(A^{-1})$ is also sufficient for the applicability of the finite section method to $A \in \text{alg } T(L^{\infty}_{N\times N})$.

3.10. Second proof of Theorem 3.7. Suppose $T(a) + K$ and $T(\tilde{a})$ are invertible on l^2_N. Then $a \in G(C + H^\infty)_{N\times N}$ by Theorem 2.35 and $T(a^{-1}) \in G\mathcal{B}(l^2_N)$ by Proposition 2.34. By virtue of Proposition 2.7 and Theorem 2.16

$$(T(a) + K)^{-1} = T(a^{-1}) + M, \quad M \in \mathfrak{C}_\infty(l^2_N) \tag{16}$$

and therefore,

$$Q_n(T(a) + K)^{-1}Q_n = Q_nT(a^{-1})Q_n + Q_nMQ_n. \tag{17}$$

Since $Q_nT(a^{-1})Q_n : \operatorname{Im} Q_n \to \operatorname{Im} Q_n$ and $T(a^{-1}) : l^2_N \to l^2_N$ are unitarily equivalent, we get the invertibility of the first operator for all n and

$$\|(Q_nT(a^{-1})Q_n)^{-1}Q_n\| = \|T^{-1}(a^{-1})\|.$$

From 1.3(d) it follows that $\|Q_nMQ_n\| \to 0$ as $n \to \infty$, which implies the invertibility of the operators (17) and the uniform boundedness of their inverses for all sufficiently large n. To conclude that $T(a) + K \in \Pi_2\{P_n\}$, it remains to apply Proposition 3.8.
A computation based on Neumann's series expansion shows that 3.7(8) is an asymptotic inverse for $P_n(T(a) + K)P_n$.
The necessity part of Theorem 3.7 is an immediate consequence of the corollary in 3.9. Δ

The local theory presented in the following enables us to give a foundation of the finite section method for Toeplitz operators with arbitrary piecewise continuous symbol, once we have obtained some information about symbols with only one discontinuity. A simple example of such a symbol is the φ^α_τ defined by 2.28(12).

3.11. Proposition (Verbickiĭ-Krupnik). Let $a = \varphi^\alpha_\tau$ and assume $1 < p < \infty$. If $T(a) \in G\mathcal{B}(l^p)$ and $T(\tilde{a}) \in G\mathcal{B}(l^p)$, then $T(a) \in \Pi_p\{P_n\}$.

Proof. Recall step 1 of the proof outline of the theorem in 2.31, according to which

$$Q_nT^{-1}(a)Q_n = c\ Q_nLT(b)MQ_n, \tag{18}$$

where $b = \varphi^{(-\alpha)}_\tau$ and where L and M are certain diagonal operators. But (18) regarded as acting on $\operatorname{Im} Q_n$, and $c\ L_nT(b)M_n : l^p \to l^p$, where

$$L_n = \operatorname{diag}\{\lambda_{n+1}, \lambda_{n+2}, \dots\}, \quad M_n = \operatorname{diag}\{\mu_{n+1}, \mu_{n+2}, \dots\},$$

are either simultaneously invertible or not.

Now

$$(L_nT(b)M_n)^{-1} = M_n^{-1}T^{-1}(b)L_n^{-1} = (1/c)M_n^{-1}MT(a)LL_n^{-1}. \qquad (19)$$

Put $\beta = \text{Re}\,\alpha$. It is not hard to verify that

$$\delta^{-1}(k+1)^{\beta} \leqq |\lambda_k| \leqq \delta(k+1)^{\beta},$$
$$\delta^{-1}(k+1)^{-\beta} \leqq |\mu_k| \leqq \delta(k+1)^{\beta}$$

with some $\delta > 0$ for $k = 0,1,2,\ldots$. Therefore,

$$\|M_n^{-1}MT(a)LL_n^{-1}\|_{\mathcal{B}(l^p)}$$

$$\leqq \sup_k |(\mu_k\lambda_k)/(\mu_{n+k}\lambda_{n+k})| \; \|M_n^{-1}MT(a)(M_n^{-1}M)^{-1}\|_{\mathcal{B}(l^p)} \qquad (20)$$

$$\leqq \delta^4\|M_n^{-1}MT(a)(M_n^{-1}M)^{-1}\|_p = \delta^4\|T(a)\|_{\mathcal{B}(l^{p,r_n})},$$

where l^{p,r_n} is the weighted l^p-space

$$l^{p,r_n} = \{\varphi = \{\varphi_k\}_{k=0}^{\infty} : \|\varphi\|_{p,r_n}^p = \sum_{k=0}^{\infty} |\varphi_k|^p\left(\frac{n+k+1}{k+1}\right)^{\beta p} < \infty\}.$$

Without loss of generality assume $\beta \geqq 0$; otherwise consider adjoints. From

$$\max\{(k+1)^{\beta p},(n+1)^{\beta p}\} \leqq (n+k+1)^{\beta p} \leqq \gamma\,((k+1)^{\beta p}+(n+1)^{\beta p})$$

we get, on the one hand,

$$\|T(a)\varphi\|_{p,r_n}^p \leqq \gamma\,(\|T(a)\varphi\|_{p,-\beta}^p(n+1)^{\beta p} + \|T(a)\varphi\|_p^p)$$

and, on the other hand,

$$\max\{\|\varphi\|_{p,-\beta}^p(n+1)^{\beta p},\ \|\varphi\|_p^p\} \leqq \|\varphi\|_{p,r_n}^p$$

(here $\|\psi\|_{p,-\beta}^p = \sum_{k=0}^{\infty} |\psi_k|^p(k+1)^{-\beta p}$, see also 4.13).

Hence

$$\|T(a)\|_{\mathcal{B}(l^{p,r_n})} \leqq \gamma^{1/p}(\|T(a)\|_{\mathcal{B}(l^{p,-\beta})} + \|T(a)\|_{\mathcal{B}(l^p)}). \qquad (21)$$

Now, the invertibility of $T(\tilde{a})$ implies $-1/p < \beta < 1/q$ by the theorem in 2.31, and this gives $T(a) \in \mathcal{B}(l^{p,-\beta})$ (see Dudučava [1], [3]). Combining (19), (20), (21) and the corollary in 3.8 we get the assertion. Δ

Remark. Note that in accordance with 2.31

$$T(\varphi_\tau^\alpha) \in G\mathcal{B}(l^p),\ T(\widetilde{\varphi}_\tau^\alpha) \in G\mathcal{B}(l^p) \iff |\mathrm{Re}\,\alpha| < \min\{\tfrac{1}{p}, \tfrac{1}{q}\},$$

where $1/p + 1/q = 1$.

LOCAL THEORY OF THE FINITE SECTION METHOD

3.12. For a Toeplitz operator to be Fredholm means to have an inverse modulo the ideal $\mathfrak{C}_\infty$ of compact operators. Now we have the formula (see 2.7)

$$T(a)T(b) = T(ab) - H(a)H(\widetilde{b}) \qquad (22)$$

and there are many cases where the occuring product of Hankel operators is known to be compact. Then $T(a)T(b)$ equals $T(ab)$ modulo the ideal $\mathfrak{C}_\infty$ and from this fact, at long last, it follows all we know about Fredholmness of Toeplitz operators. In particular, the applicability of the local principle of Gohberg and Krupnik essentially rests on the identity (22) and on the fact, that $H(f) \in \mathfrak{C}_\infty$ for $f \in C$.

The analogue of (22) for finite Toeplitz matrices reads (see 3.6):

$$T_n(a)T_n(b) = T_n(ab) - P_nH(a)H(\widetilde{b})P_n - W_nH(\widetilde{a})H(b)W_n. \qquad (23)$$

According to Proposition 3.3 the applicability of the finite section method to $T(a)$ is equivalent to the existence of an inverse $\{R_n\} \in \mathcal{F}_c$ for $\{T_n(a)\} \in \mathcal{F}_c$ modulo the ideal $\mathfrak{N}$. Looking for a connection between $\mathfrak{N}$ and (23) we observe that the ideal $\mathfrak{N}$ is, in a certain sense, too small; so sequences of the form

$$\{P_nKP_n + W_n\widetilde{K}W_n\}, \quad \text{where } K, \widetilde{K} \in \mathfrak{C}_\infty, \qquad (24)$$

do, in general, not belong to $\mathfrak{N}$. Thus, in order to develop a theory of the finite section method in analogy to the Fredholm theory, it would be desirable to have an ideal $\mathfrak{J}$ that contains all sequences of the form (24). But there is no such ideal in $\mathcal{F}_c$ – this algebra is, in a certain sense, too large. In what follows we shall construct a somewhat smaller algebra $\mathfrak{A}$ possessing, on the one hand, such an ideal and containing, on the other hand, sufficiently much interesting elements, in particular, all sequences of the form $\{T_n(a)\}$.

3.13. Definitions. For $B \in \mathcal{B}(l^p_N)$ $(1 < p < \infty)$, we denote by $B^* \in \mathcal{B}(l^q_N)$ $(1/p + 1/q = 1)$ the adjoint operator. Let P_n and W_n denote the operators introduced in 3.4 and 3.5. Given $A_n : \mathrm{Im}\, P_n \to \mathrm{Im}\, P_n$ we set

$$\widetilde{A}_n := W_nA_nW_n : \mathrm{Im}\, P_n \to \mathrm{Im}\, P_n.$$

Now define $\mathfrak{A}$ to be the subset of $\mathcal{F}$ consisting of all sequences $\{A_n\}_{n=0}^{\infty}$ such that there exist $A \in \mathcal{B}(l_N^p)$ and $\tilde{A} \in \mathcal{B}(l_N^p)$ satisfying

$$A_n P_n \to A, \quad A_n^* P_n \to A^*$$

$$\tilde{A}_n P_n \to \tilde{A}, \quad \tilde{A}_n^* P_n \to \tilde{A}^*$$

(recall that $\to$ denotes strong convergence).
It is easy to see that $\mathfrak{A}$ forms a closed subalgebra of $\mathcal{F}$ (even of $\mathcal{F}_c$), that is, $\mathfrak{A}$ itself is a Banach algebra.

Obviously, $\mathfrak{N} \subset \mathfrak{A}$. We have $\{P_n K P_n\} \in \mathfrak{A}$ for every $K \in \mathfrak{C}_\infty(l_N^p)$ due to 1.3(d). It is easy to deduce from 1.1(d) that also $\{W_n K W_n\} \in \mathfrak{A}$ for all $K \in \mathfrak{C}_\infty(l_N^p)$.

Finally, the identity $W_n T_n(a) W_n = T_n(\tilde{a})$ implies that $\{T_n(a)\} \in \mathfrak{A}$ for every $a \in (M_p)_{N\times N}$. Moreover, it can be shown that even $\{P_n A P_n\} \in \mathfrak{A}$ if A is an arbitrary operator in $\operatorname{alg} T((M_p)_{N\times N})$ and that in this case $\tilde{A} = \widetilde{S(A)}$ (cf. Roch, Silbermann [2]).

Now, denote by $\mathfrak{J}$ the collection of all sequences $\{A_n\} \in \mathcal{F}$ of the form

$$A_n = P_n K P_n + W_n \tilde{K} W_n + C_n,$$

where $K \in \mathfrak{C}_\infty(l_N^p)$, $\tilde{K} \in \mathfrak{C}_\infty(l_N^p)$ and $\{C_n\} \in \mathfrak{N}$.

<u>3.14. Proposition.</u> $\mathfrak{J}$ is a two-sided closed ideal in $\mathfrak{A}$.

<u>Proof.</u> Obviously, $\mathfrak{J}$ is a linear subspace of $\mathfrak{A}$. Let us prove that $\mathfrak{J}$ is closed. If

$$\{A_n\} = \{P_n K P_n + W_n \tilde{K} W_n + C_n\} \in \mathfrak{J},$$

then $A_n \to K$ and $\tilde{A}_n \to \tilde{K}$, therefore

$$\|K\| < \liminf \|A_n\|, \quad \|\tilde{K}\| \leqq \liminf \|\tilde{A}_n\|.$$

Thus if $\{A_n^{(j)}\} \subset \mathfrak{J}$ is a Cauchy sequence, then so are $\{K^{(j)}\}$ and $\{\tilde{K}^{(j)}\}$ in $\mathfrak{C}_\infty(l_N^p)$, consequently, there exist compact operators K and $\tilde{K}$ such that $K^{(j)} \rightrightarrows K$ and $\tilde{K}^{(j)} \rightrightarrows \tilde{K}$. Now it follows almost at once that there is a sequence $\{A_n\} \in \mathfrak{J}$ such that $\|\{A_n^{(j)}\} - \{A_n\}\|_{\mathfrak{A}} \to 0$ as $j \to \infty$.

Finally, let $\{A_n\} \in \mathfrak{J}$ and $\{B_n\} \in \mathfrak{A}$. Then

$$\begin{aligned} B_n A_n &= B_n P_n K P_n + B_n W_n \tilde{K} W_n + B_n C_n \\ &= P_n B_n K P_n + W_n \tilde{B}_n P_n \tilde{K} W_n + B_n C_n \\ &= P_n B K P_n + W_n \tilde{B} \tilde{K} W_n + C_n', \end{aligned}$$

where $\{C_n'\} \in \mathfrak{N}$. Analogously can be proved that $\{A_nB_n\} \in \mathfrak{J}$. Δ

3.15. Definition. We put $\mathfrak{A}^\circ = \mathfrak{A}/\mathfrak{J}$ and denote by $\{A_n\}^\circ \in \mathfrak{A}^\circ$ the coset containing $\{A_n\} \in \mathfrak{A}$.

3.16. Theorem. Let $A \in \mathfrak{B}(l_N^p)$, $1 < p < \infty$, and let $\{A_n\} \in \mathfrak{A}$ be a sequence such that $A_nP_n \to A$. Then

$$A \in \Pi\{A_n\} \iff A \in G\mathfrak{B}(l_N^p),\ \ \mathfrak{X} \in G\mathfrak{B}(l_N^p) \text{ and } \{A_n\}^\circ \in G\mathfrak{A}^\circ$$

(here, of course, $\mathfrak{X} = \underset{n\to\infty}{\text{s-lim}}\, W_nA_nW_n$).

Thereby, if $\{R_n\}^\circ$ is the inverse of $\{A_n\}^\circ$ in $\mathfrak{A}^\circ$, then

$$\pi_c\{R_n + P_n(A^{-1} - R)P_n + W_n(\mathfrak{X}^{-1} - \mathfrak{R})W_n\} \qquad (25)$$

is the inverse of $\pi_c\{A_n\}$ in $\mathfrak{F}_c/\mathfrak{N}$ (here $R = \underset{n\to\infty}{\text{s-lim}}\, R_nP_n$, $\mathfrak{R} = \underset{n\to\infty}{\text{s-lim}}\, W_nR_nW_n$).

Proof. Suppose $A \in \Pi\{A_n\}$. Then $A \in G\mathfrak{B}(l_N^p)$ by Proposition 3.3. Let us prove that $\mathfrak{X} \in G\mathfrak{B}(l_N^p)$. Again from Proposition 3.3 we may conclude that $\mathfrak{X}_n = W_nA_nW_n$ is invertible for all sufficiently large n and

$$\sup_n \|\mathfrak{X}_n^{-1}P_n\| = \sup_n \|W_nA_n^{-1}W_n\| = \sup_n \|A_n^{-1}P_n\| =: C < \infty . \qquad (26)$$

Thus $\|P_n\varphi\| \leqq C\|\mathfrak{X}_nP_n\varphi\|$ for all $\varphi \in l_N^p$, and passing to the limit $n \to \infty$ we get $\|\varphi\| \leqq C\|\mathfrak{X}\varphi\|$ for all $\varphi \in l_N^p$. Analogously can be shown that $\|\psi\| \leqq C\|\mathfrak{X}^*\psi\|$ for all $\psi \in l_N^q = (l_N^p)^*$. Hence $\mathfrak{X} \in G\mathfrak{B}(l_N^p)$.
Now it follows that $A^* \in G\mathfrak{B}(l_N^q)$ and $\mathfrak{X}^* \in G\mathfrak{B}(l_N^q)$ and in view of (26) and Proposition 3.3 we get

$$\mathfrak{X} \in \Pi\{\mathfrak{X}_n\},\quad A^* \in \Pi\{A_n^*\},\quad \mathfrak{X}^* \in \Pi\{\mathfrak{X}_n^*\}.$$

Let $\{B_n\} \in \mathfrak{F}_c$ be the inverse of $\{A_n\} \in \mathfrak{F}_c$ modulo $\mathfrak{N}$, that is, $B_nA_n = P_n + C_n$, $\{C_n\} \in \mathfrak{N}$. Then

$$\mathfrak{B}_n = (P_n + W_nC_nW_n)\,\mathfrak{X}_n^{-1}P_n \to \mathfrak{X}^{-1},$$

$$B_n^* = (A_n^*)^{-1}(P_n + C_n^*)P_n \to (A^*)^{-1},$$

$$\mathfrak{B}_n^* = (\mathfrak{X}_n^*)^{-1}(P_n + W_nC_n^*W_n)P_n \to (\mathfrak{X}^*)^{-1}.$$

Therefore $\{B_n\} \in \mathfrak{A}$ and, consequently, $\{B_n\}^\circ$ is the inverse of $\{A_n\}^\circ$ in $\mathfrak{A}^\circ$, i.e. $\{A_n\}^\circ \in G\mathfrak{A}^\circ$.
Conversely, suppose now A and $\mathfrak{X}$ are invertible in $\mathfrak{B}(l_N^p)$ and $\{A_n\}^\circ$ is so in $\mathfrak{A}^\circ$. Then there is a sequence $\{R_n\} \in \mathfrak{A}$ such that

$$A_nR_n = P_n + P_nKP_n + W_n\tilde{K}W_n + C_n,$$

where K and $\tilde{K}$ are compact and $\{C_n\} \in \mathcal{N}$. Passage to the limit $n \to \infty$ gives

$$AR = I + K, \quad \tilde{A}\tilde{R} = I + \tilde{K}.$$

Thus, $A^{-1} - R =: T \in \mathfrak{C}_\infty(l^p_N)$, $\tilde{A}^{-1} - \tilde{R} =: \tilde{T} \in \mathfrak{C}_\infty(l^p_N)$.

Put $R'_n = R_n + P_nTP_n + W_n\tilde{T}W_n$. Then, obviously, $\{R'_n\} \in \mathcal{A}$ and

$$A_nR'_n = P_n + P_n(K + A_nP_nT)P_n + W_n(\tilde{K} + \tilde{A}_nP_n\tilde{T})W_n + C_n$$

$$= P_n + P_n(K + AT)P_n + W_n(\tilde{K} + \tilde{A}\tilde{T})W_n + C'_n$$

$$= P_n + C'_n,$$

where $\{C'_n\} \in \mathcal{N}$. In the same way we may show that $R'_nA_n = P_n + C''_n$, $\{C''_n\} \in \mathcal{N}$. This gives the remaining assertions of our theorem. Δ

3.17. Third proof of Theorem 3.7. We apply Theorem 3.16 with $A = T(a) + K$ and $A_n = P_n(T(a) + K)P_n$. Then, if $T(a) + K \in \Pi\{P_n\}$, we get at once the invertibility of $T(a) + K$ and that of

$$T(\tilde{a}) = \operatorname*{s-lim}_{n\to\infty} W_n(T(a) + K)W_n$$

in $\mathcal{B}(l^2_N)$. Conversely, suppose $T(a) + K$ and $T(\tilde{a})$ are invertible. It remains to show that then $\{P_n(T(a) + K)P_n\}^o \in G\mathcal{A}^o$. To that end notice that $a \in G(C + H^\infty)$ by Theorem 2.35, therefore $H(\tilde{a}), H(\widetilde{a^{-1}}) \in \mathfrak{C}_\infty(l^2_N)$ by Theorem 2.16 and, consequently, by Propositions 3.6 and 3.14

$$\{P_n(T(a) + K)P_n\}^o\{T_n(a^{-1})\}^o$$

$$= \{P_n\}^o - \{P_nH(a)H(\widetilde{a^{-1}})P_n\}^o - \{W_nH(\tilde{a})H(a^{-1})W_n\}^o$$

$$+ \{P_nKP_nT_n(a^{-1})\}^o$$

$$= \{P_n\}^o .$$

Equally, $\{T_n(a^{-1})\}^o\{P_n(T(a) + K)P_n\}^o = \{P_n\}^o$. From 3.16(25) we deduce that 3.7.(8) is an inverse of $P_n(T(a) + K)P_n$ modulo $\mathcal{N}$. Δ

Remark. By virtue of Theorem 2.21 and Proposition 2.20 the same arguments apply to get that for $a \in (C \cap M_{<p>})_{N\times N}$ and $K \in \mathfrak{C}_\infty(l^p_N)$

$$T(a) + K \in \Pi_p\{P_n\} \iff T(a)+K \in G\mathcal{B}(l^p_N),\ T(\tilde{a}) \in G\mathcal{B}(l^p_N).$$

3.18. Proposition. Let $a,b \in M_p$ and suppose

$$T(a) \in \Pi_p\{P_n\}, \quad T(b) \in \Pi_p\{P_n\},$$

$$H(a)H(\tilde{b}) \in \mathfrak{C}_\infty(l^p), \quad H(\tilde{a})H(b) \in \mathfrak{C}_\infty(l^p).$$

Then $T(ab) \in \Pi_p\{P_n\}$.

Proof. From $T(ab) = T(a)T(b) + H(a)H(\tilde{b})$ it follows that $T(ab) \in \Phi(l^p)$ and $\operatorname{Ind} T(ab) = 0$. Thus, by the corollary in 2.8, the operator $T(ab)$ is invertible. Analogously one can prove the invertibility of $T(\widetilde{ab})$. Lastly, $\{T_n(ab)\}^o = \{T_n(a)\}^o\{T_n(b)\}^o$ by Proposition 3.6 and it remains to apply Theorem 3.16. Δ

We denote by sing supp a the complement of the largest open subset of Γ on which $a \in L^\infty$ is continuous. From Proposition 2.24 we know that $H(a)H(\tilde{b}) \in \mathfrak{C}_\infty(l^p)$ if $a,b \in M_{<p>}$ and

$$\text{sing supp } a \cap \text{sing supp } b = \emptyset.$$

Thus the above theorem has the following corollary.

Corollary 1. Let $a_1,\dots,a_R \in M_{<p>}$ and suppose $T(a_r) \in \Pi_p\{P_n\}$ $(r = 1,\dots,R)$. If

$$\text{sing supp } a_r \cap \text{sing supp } a_s = \emptyset \quad (r \neq s)$$

then $T(a_1 \dots a_R) \in \Pi_p\{P_n\}$.

Finally, combining the above theorem with 2.28, 2.31, and Proposition 3.11 we get the following.

Corollary 2. Let $a \in PC \cap M_{<p>}$. Then $T(a) \in \Pi_p\{P_n\}$ if and only if $T(a)$ and $T(\tilde{a})$ are invertible in $\mathfrak{B}(l^p)$.

The extension of the latter result to symbols in $\overline{PC} \cap M_{<p>}$ will be our next concern.

3.19. Theorem. Let $a \in \overline{PC} \cap M_{<p>}$ and $K \in \mathfrak{C}_\infty(l^p)$. Then $T(a) + K \in \Pi_p\{P_n\}$ if and only if $T(a) + K$ and $T(\tilde{a})$ are invertible in $\mathfrak{B}(l^p)$.

Proof. The necessity of the conditions immediately results from Theorem 3.16.

To prove their sufficiency we apply the local principle of Gohberg and Krupnik to the algebra $\mathfrak{A}^o$.

For $\tau = e^{i\vartheta} \in \Gamma$ define N_τ as in the proof of Theorem 2.21 and set

$$N_\tau^o = \{\{T_n(f)\}^o \in \mathfrak{A}^o : f \in N_\tau\}.$$

It is a simple application of Proposition 3.6 (see also the remark in 3.17) to show that $\{N_\tau^o\}_{\tau\in\Gamma}$ forms a covering system of localizing classes in $\mathfrak{A}^o$ and that $\{P_n(T(a) + K)P_n\}^o = \{T_n(a)\}^o$ belongs to the commutant of $\bigcup_{\tau\in\Gamma} N_\tau^o$.

Now fix $\tau \in \Gamma$. If $a(t)$ is continuous at τ then define a_τ as $a_\tau(t) \equiv a(\tau)$. In case $a(\tau - 0) \neq a(\tau + 0)$ choose an $a_\tau \in M_{<p>}$ satisfying

$$a_\tau(\tau) = a(\tau), \quad a_\tau(\tau \pm 0) = a(\tau \pm 0)$$

and being continuous on $\Gamma \setminus \{\tau\}$; the function a_τ will be more specified below. Then, first of all, $\{T_n(a)\}^o$ is N_τ^o-equivalent to $\{T_n(a_\tau)\}^o$. Indeed, the infimum taken over $f \in N_\tau$,

$$\begin{aligned}\inf\|\{T_n(a - a_\tau)T_n(f)\}^o\| &= \inf\|\{T_n[(a - a_\tau)f]\}^o\| \\ &\leqq \inf\|(a - a_\tau)f\|_{M_p}\end{aligned}$$

and if r is chosen in a neighborhood of p such that $a, a_\tau \in M_r$, then by 2.6(d)

$$\begin{aligned}&\inf\|(a - a_\tau)f\|_{M_p} \\ &\leqq \inf\|(a - a_\tau)f\|_{L^\infty}^{1-\gamma}\|(a - a_\tau)f\|_{M_r}^{\gamma}, \quad \gamma = \frac{r|p-2|}{p|r-2|},\end{aligned}$$

and this becomes 0 since $0 < \gamma < 1$,

$$\inf\|(a - a_\tau)f\|_{L^\infty} = 0,$$

and since by virtue of 2.2(5)

$$\begin{aligned}&\|(a - a_\tau)f\|_{M_r} \leqq \|a - a_\tau\|_{M_r}\|f\|_{M_r} \\ &\leqq \|a - a_\tau\|_{M_r} \cdot c_r\{\|f\|_{L^\infty} + V_1(f)\} = 3c_r\|a - a_\tau\|_{M_r}.\end{aligned}$$

Now the invertibility of $T(a) + K$ implies that of $T(a)$ by the corollary in 2.8. From the invertibility of $T(a)$ and $T(\tilde{a})$ by means of a geometrical consideration based on the facts in 2.31 one may conclude that the origin must not ly in the (closed) lentiform domain having as its boundary the circular arcs

$$\{(1 - s_p(\mu))a(\tau - 0) + s_p(\mu)a(\tau + 0) : \mu \in [0,1]\},$$

$$\{(1 - s_p(\mu))a(\tau + 0) + s_p(\mu)a(\tau - 0) : \mu \in [0,1]\}.$$

This implies that a_τ can be chosen in such a way that both $T(a_\tau)$ and $T(\tilde{a}_\tau)$ are invertible in $\mathcal{B}(l^p)$. Then we get $T(a_\tau) \in \Pi_p\{P_n\}$ from corollary in 3.18, and therefore $\{T_n(a_\tau)\}^o \in G\mathfrak{A}^o$ by Theorem 3.16. To complete the proof it remains to apply the theorem in 1.14. Δ

Before generalizing Theorem 3.19 to matrix symbols we need a fact which is almost obvious in the scalar case (and has already been used, when we were proving Theorem 3.19) but requires a proof in the matrix case.

3.20. Lemma. Let $a \in (PC \cap M_{<p>})_{N\times N}$, $1 < p < \infty$. For $T(a) \in \Phi(l_N^v)$ denote by $\mathrm{Ind}_v T(a)$ the index of $T(a)$ on l_N^v. Suppose

$$\min\{p,q\} \leqq r_1 \leqq r_2 \leqq \max\{p,q\}, \text{ where } 1/p + 1/q = 1.$$

If $T(a) \in \Phi(l_N^{r_1})$ and $T(a) \in \Phi(l_N^{r_2})$ then

$$\mathrm{Ind}_{r_1} T(a) \leqq \mathrm{Ind}_{r_2} T(a), \tag{27}$$

and equality holds if and only if $T(a) \in \Phi(l_N^s)$ for all $s \in [r_1, r_2]$.

Proof. Let $a = b\varphi c$ be the factorization in accordance with Lemma 2.37. Then

$$\det[a_s(t,\mu)] = \det b(t) \cdot \det[\varphi_s(t,\mu)] \cdot \det c(t) \tag{28}$$

and denoting by φ_k the diagonal entries of φ,

$$\det[\varphi_s(t,\mu)] = \prod_{k=1}^{N} (\varphi_k)_s(t,\mu). \tag{29}$$

By our hypotheses, b and c are regular matrix functions, hence for $i = 1,2$ and $k = 1,\dots,N$

$$(\varphi_k)_{r_i}(t,\mu) \neq 0, \quad \mathrm{ind} \prod_{k=1}^{N} (\varphi_k)_{r_i}(t,\mu) = \varkappa_i,$$

$$\varkappa_i = - \mathrm{Ind}_{r_i} T(a) - \mathrm{ind} \det b - \mathrm{ind} \det c. \tag{30}$$

Since the index of a product is equal to the sum of the indices of the factors, we get

$$\sum_{k=1}^{N} \mathrm{ind}(\varphi_k)_{r_1} = \varkappa_1 , \quad \sum_{k=1}^{N} \mathrm{ind}(\varphi_k)_{r_2} = \varkappa_2 . \tag{31}$$

A little geometrical consideration shows that in the scalar case for every $\psi \in PC$ satisfying $\psi_{r_1}(t,\mu) \neq 0$ and $\psi_{r_2}(t,\mu) \neq 0$ for all $(t,\mu) \in \Gamma \times [0,1]$ the relation $\mathrm{ind}\, \psi_{r_1} \leqq \mathrm{ind}\, \psi_{r_2}$ holds and that we have equality if and only if $\psi_s(t,\mu) \neq 0$ for all $s \in [r_1, r_2]$.
Thus $\varkappa_1 \leqq \varkappa_2$, and (30), (31) give (27).
If $\varkappa_1 = \varkappa_2$ then (31) and $\mathrm{ind}(\varphi_k)_{r_1} \leqq \mathrm{ind}(\varphi_k)_{r_2}$ (for all k) imply $\mathrm{ind}\,(\varphi_k)_{r_1} = \mathrm{ind}(\varphi_k)_{r_2}$ and therefore, $(\varphi_k)_s(t,\mu) \neq 0$ for all $s \in [r_1, r_2]$. From (28), (29) it follows that $T(a) \in \Phi(l_N^s)$.
Finally, if $T(a) \in \Phi(l_N^s)$ for all $s \in [r_1, r_2]$ then $\mathrm{ind}(\varphi_k)_{r_1} = \mathrm{ind}(\varphi_k)_{r_2}$

for all k and (30), (31) yield the equality $\mathrm{Ind}_{r_1}T(a) = \mathrm{Ind}_{r_2}T(a)$. Δ

Corollary. Let $a \in (PC \cap M_{<p>})_{N\times N}$, $1 < p < \infty$.
Suppose $T(a)$ and $T(\tilde{a})$ are in $\Phi(l^p_N)$ and $\mathrm{Ind}_p T(a) = \mathrm{Ind}_p T(\tilde{a}) = 0$. Then $T(a) \in \Phi(l^s_N)$ and $\mathrm{Ind}_s T(a) = 0$ for all s between p and q $(1/p + 1/q = 1)$.

Proof. By Theorem 2.38, $T(\tilde{a}) \in \Phi(l^p_N)$ if and only if $T(a) \in \Phi(l^q_N)$ and in this case $\mathrm{Ind}_p T(\tilde{a}) = -\mathrm{Ind}_q T(a)$. Δ

3.21. Theorem. Let $a \in (\overline{PC} \cap M_{<p>})_{N\times N}$, $1 < p < \infty$, and $K \in \mathfrak{C}_\infty(l^p_N)$. Then $T(a) + K \in \prod_p\{P_n\}$ if and only if $T(a) + K$ and $T(\tilde{a})$ are invertible in $\mathcal{B}(l^p_N)$.

Proof. The necessity of the conditions is again an immediate consequence of Theorem 3.16. Thus suppose $T(a) + K$ and $T(\tilde{a})$ are invertible on l^p_N. In view of Theorem 3.16 we are left with the proof of the invertibility of $\{P_n(T(a) + K)P_n\}^o = \{T_n(a)\}^o$ in $\mathfrak{A}^o$.
Define N_τ as in the proof of Theorem 2.21 and put

$$N^o_\tau = \{\{\mathrm{diag}(T_n(f),\dots,T_n(f))\}^o \in \mathfrak{A}^o : f \in N_\tau\} .$$

Then $\{N^o_\tau\}_{\tau\in\Gamma}$ forms a covering system of localizing classes in $\mathfrak{A}^o$, and $\{T_n(a)\}^o$ commutes with every element of the union of the classes N^o_τ. Define a_τ by $a_\tau(t) \equiv a(\tau)$ if $a(t)$ is continuous at τ.
In case $a(\tau - 0) \neq a(\tau + 0)$ let a_τ be any matrix function in $M_{<p>}$ such that

$$a_\tau(\tau) = a(\tau), \quad a_\tau(\tau \pm 0) = a(\tau \pm 0), \tag{32}$$

$$a_\tau \text{ is continuous on } \Gamma \setminus \{\tau\}, \tag{33}$$

$$\det[a_\tau(t \pm 0)] \neq 0 \text{ for all } t \in \Gamma . \tag{34}$$

Then, as in 3.19, the coset $\{T_n(a)\}^o$ can be proved to be N^o_τ-equivalent to $\{T_n(a_\tau)\}^o$. Let $a_\tau = b\varphi c$ be the factorization ensured by Lemma 2.37, and denote the diagonal entries of φ by φ_i $(i=1,\dots,N)$. Then above all,

$$\{T_n(a_\tau)\}^o = \{T_n(b)\}^o\{T_n(\varphi)\}^o\{T_n(c)\}^o$$

and since $\{T_n(b)\}^o$ and $\{T_n(c)\}^o$ are invertible in $\mathfrak{A}^o$ (the inverses are $\{T_n(b^{-1})\}^o$ and $\{T_n(c^{-1})\}^o$), it remains to prove the invertibility of $\{T_n(\varphi)\}^o$.
From the corollary in 3.20 we get $T(\varphi) \in \Phi(l^s_N)$ for all s between p and q $(1/p + 1/q = 1)$. Thus $T(\varphi_i) \in \Phi(l^s)$ for these s, and

again from 3.20 (with $N = 1$) it follows that there are integers m_i $(i = 1,\dots,N)$ such that $T(\chi_{m_i}\varphi_i) \in G\mathcal{B}(l^r)$ for all r between p and q. In particular, $T(\chi_{m_i}\varphi_i) \in G\mathcal{B}(l^p)$ and $T(\tilde{\chi}_{m_i}\tilde{\varphi}_i) \in G\mathcal{B}(l^p)$, which by the Theorems 3.19 and 3.16 implies the invertibility of $\{T_n(\chi_{m_i}\varphi_i)\}^o$. But

$$\{T_n(\chi_{m_i}\varphi_i)\}^o = \{T_n(\chi_{m_i})\}^o\{T_n(\varphi_i)\}^o = \{T_n(\varphi_i)\}^o\{T_n(\chi_{m_i})\}^o,$$

so that $\{T_n(\varphi_i)\}^o$ is invertible. Thus

$$\{\Phi_n\}^o := \{\mathrm{diag}(T_n(\varphi_1),\dots,T_n(\varphi_N))\}^o \in G\mathfrak{A}^o.$$

Therefore $\{T_n(\varphi)\}^o$ may be written in the form

$$\{T_n(\varphi)\}^o = \{\Phi_n\}^o\{T_n(\chi_o) + X_n\}^o,$$

where $\{X_n\} \in \mathfrak{A}$ has the property that $\{X_n^N\}^o = \{0\}^o$. As a consequence of this, $\{T_n(\chi_o) + X_n\}^o \in G\mathfrak{A}^o$, since if $x^N = 0$ then

$$\begin{aligned} e &= (e + x - x)^N \\ &= (-1)^N x^N + \sum_{k=1}^{N} \binom{N}{k}(e + x)^k(-x)^{N-k} \\ &= (e + x) \sum_{k=1}^{N} \binom{N}{k}(e + x)^{k-1}(-x)^{N-k} \\ &= \Big(\sum_{k=1}^{N} \binom{N}{k}(e + x)^{k-1}(-x)^{N-k}\Big)(e+x), \end{aligned}$$

which yields the invertibility of $e + x$. Δ

NOTES AND COMMENTS

General references: Gohberg, Feldman [1], Prößdorf [1], Prößdorf, Silbermann [1].

3.3. See Gohberg, Feldman [1]. The equivalence

$$A \in \Pi\{A_n\} \iff A \in G\mathcal{B}(X), \quad \pi\{A_n\} \in G(\mathcal{F}/\mathcal{N})$$

inspired Kozak to the utilization of local principles to the investigation of the finite section method, see Kozak [1], [2]. He applied his local principle, which was originated by that of Simonenko, to the algebra $\mathcal{F}/\mathcal{N}$ and obtained a series of remarkable results on higher-dimensional convolution operators with continuous symbols. The local method presented here is based on the equivalence (see 3.16)

$$A \in \Pi\{A_n\} \iff A \in G\mathcal{B}(X),\ \tilde{A} \in G\mathcal{B}(X),\ \{A_n\}^o \in G(\mathfrak{A}/\mathfrak{z})$$

($\{A_n\} \in \mathfrak{A}$). Note that the algebra $\mathfrak{A}/\mathfrak{J}$ is considerably better fitted to our purposes than $\mathfrak{F}/\mathfrak{N}$. However, there are algebras smaller than $\mathfrak{A}/\mathfrak{J}$ of much better structure (see Böttcher, Silbermann [4]).

3.6. Although similar relations have been used for a long time, the identity (4) and the resulting formula (5) explicitly appeared in Widom [8] for the first time.

3.7. This theorem (for $K = 0$) was first proved by Ambarcumian [1] and subsequently several proofs were found (one also contained in Widom [8]). Each of the three proofs presented in the text is of its own interest and allows extensions to certain other situations.
The proof given in section 3.7 is published here for the first time (but see also Kozak, Simonenko [1]). It can be used to derive e.g. the following result (see Böttcher [2]). Put

$$l^{2,\infty} = \bigcap_{\mu \geqq 0} l_2^{\mu}, \quad l^{2,-\infty} = \bigcup_{\mu \geqq 0} l_2^{-\mu}$$

(cf. 4.13), equip the first space with a countable-norm topology (see Prößdorf [1], chapter 9), think of the second as the dual of the first and provide it with the strong topology. Then for $a \in \tilde{L}^{2,\infty}_{N\times N}$, $K \in \mathfrak{C}_\infty(l^{2,\infty}_N)$ (in the sense of the lemma on p.312 of Prößdorf [1]) the finite section method is applicable to $T(a) + K$ in $l^{2,\infty}_N$ if and only if $T(a) + K$ is invertible on $l^{2,\infty}_N$ and $T(\tilde{a})$ is so on $l^{2,-\infty}_N$. Note that in this way degenerate (non-elliptic) symbols can be included into the consideration.

3.8. This proposition was communicated to one of the authors by Kozak. In Silbermann [1] its role for the study of Toeplitz determinants was recognized and so it was published there for the first time. Since this time it has been published in several places: Silbermann, Rost [1], Silbermann [2], Böttcher, Silbermann [1], [2], Kozak, Simonenko [1].

3.9. These results are published here for the first time and were suggested by Roch, Silbermann [2].

3.10. See Böttcher, Silbermann [1]. For extension of this approach to other directions see Silbermann, Rost [1] and Silbermann [5] (Integro-difference equations).

3.11. See Verbickiĭ, Krupnik [1], where even l^p-spaces with weight are considered.

3.12. - 3.21. The local theory of the finite section method presented in the text is due to Silbermann [6]. The note Böttcher, Silbermann [3] containing the results of 3.18 must be regarded as a forerunner of the paper Silbermann [6]; it was written some months before the latter

paper, but appeared only after this work.
Theorem 3.19 was already proved in Gohberg, Feldman [1] for the case $p = 2$ if the symbol has only finitely many jumps. Then Verbickiĭ and Krupnik [1] showed its necessity part and succeeded in verifying the sufficiency portion for the case of one discontinuity. Later this result turned out to be of great importance, since it solves the corresponding problem for the local representatives arising in the theory of Silbermann [6]. The case of finitely many discontinuities was covered in Böttcher, Silbermann [3], but the final solution (including also countably many jumps, block Toeplitz operators and compact perturbations of them) was given in Silbermann [6] only recently.
Lemma 3.20 and the proofs of the theorems 3.19 and 3.21 are published here for the first time, although basic ideas go back to the case $p = 2$ presented in Silbermann [6].

In connection with Theorem 3.16 we mention the following conjecture: if $a \in L^\infty$ then

$$T(a) \in \Phi(H^2) \iff \{T_n(a)\}^o \in G\mathfrak{A}^o.$$

Clearly, the confirmation of this conjecture would completely solve the problem of the applicability of the finite section method to $T(a)$ for arbitrary $a \in L^\infty$. It is obvious from the definition of $\mathfrak{A}^o$ that the implication " $\Leftarrow$ " is true for general $a \in L^\infty$.
On the other hand, the reverse implication " $\Rightarrow$ " easily follows for $a \in C + H^\infty$ or $a \in \overline{PC}$ from the Theorems 3.7 and 3.19.

For extensions of the local method developed here to other situations we refer to the following papers.

Roch, Silbermann [1] (Powers of Toeplitz operators):

If $a \in \overline{PC} \cap M_{<p>}$ and $k \in \mathbb{Z}_+$, then $T^k(a) \in \Pi_p\{P_n\}$ if and only if $T^k(a)$ and $T(\tilde{a}^k)$ both are invertible on l^p.

Roch, Silbermann [2] (Weighted l^p-spaces):

If $a \in \overline{PC} \cap M_{<p,\gamma>}$ (cf. 4.13), $-1/p < \gamma < 1/q$, then $T(a) \in \Pi_{p,\gamma}\{P_n\}$ if and only if $T(a) \in G\mathcal{B}(l^\gamma_p)$ and $T(\tilde{a}) \in G\mathcal{B}(l^p)$; equally for the block case.

Silbermann [7] (Paired convolutions and singular integral operators):

(a) Define $\mathcal{P}_n$ on L^2 as

$$\mathcal{P}_n\Big(\sum_{k=-\infty}^{\infty} f_k t^k\Big) = \sum_{k=-n}^{n} f_k t^k \quad (|t| = 1)$$

and let P and Q denote the usual projections on L^2. If $a, b \in \overline{PC}$ and if in addition

$$\left| \arg \frac{a(t+0)}{a(t-0)} - \arg \frac{b(t+0)}{b(t-0)} \right| < \pi \quad \text{for every} \quad t \in \Gamma$$

(with $\arg z \in (-\pi,\pi)$ for $z \in \mathbb{C}$, $z \neq 0$), then

$$M(a)P + M(b)Q \in \Pi_{L^2}\{\tilde{P}_n(M(a)P + M(b)Q)\tilde{P}_n\}$$

$$\iff M(a)P + M(b)Q,\ M(a)P + Q,\ P + M(b)Q \in G\mathcal{B}(L^2)$$

(for $a,b \in PC$ this was already proved in Verbickiĭ [3]).

(b) If $a,b \in \overline{PC} \cap M_{<p>}$ have no common points of discontinuity, then

$$M(a)P + M(b)Q \in \Pi_{\ell^p}\{\tilde{P}_n(M(a)P + M(b)Q)\tilde{P}_n\}$$

$$\iff M(a)P + M(b)Q,\ M(\tilde{a})P + Q,\ P + M(\tilde{b})Q \in G\mathcal{B}(\ell^p)$$

(the above definition of $\tilde{P}_n$ extends from $L^2 \cong \ell^2$ to ℓ^p in an obious way).

(c) If $a,b \in (\overline{PC})_{N\times N}$ have no common points of discontinuity, then

$$M(a)P + M(b)Q \in \Pi_{L^2_N}\{\tilde{P}_n(M(a)P + M(b)Q)\tilde{P}_n\}$$

$$\iff M(a)P + M(b)Q,\ M(\tilde{a})P + Q,\ P + M(\tilde{b})Q \in G\mathcal{B}(L^2_N).$$

Analogous results hold for the operators $PM(a) + QM(b)$.

Fortunately in each of these cases the local representatives were already investigated again by Verbickiĭ [2], [3] and Verbickiĭ, Krupnik [1].
See also Böttcher [3], [5] (Wiener-Hopf integral operators), Junghanns, Silbermann [1] (collocation for singular integral operators), Meyer [1] (singular integral operators with certain Carleman shifts).

Wiener-Hopf integral operators. Almost all problems considered hitherto for (discrete) Toeplitz operators arise for their continuous analogue, the Wiener-Hopf integral operators, as well. For lack of space we restrict ourselves to give only some informations.

For $a \in L^\infty(\mathbb{R})$ the Wiener-Hopf integral operator on $L^2(\mathbb{R}_+)$ with the symbol $a(\xi)$ is defined as

$$W(a) : L^2(\mathbb{R}_+) \to L^2(\mathbb{R}_+), \quad \varphi \mapsto PF^{-1}a(\cdot)F\varphi,$$

where F denotes the Fourier-Plancherel operator, F^{-1} its inverse, and P acts by the rule

$$(P\varphi)(x) = (1/2 + 1/2 \cdot \operatorname{sign} x)\varphi(x).$$

The classes introduced in 2.2, 2.19 and 2.42 have their continuous analogue and so Wiener-Hopf integral operators on $L^p(\mathbb{R}_+)$ $(1 \leq p < \infty)$ may also be defined. If $a(\xi)$ is of the form $a = c + Fk$, where $c \in \mathbb{C}$, $k \in L^1(\mathbb{R})$, then $W(a)$ is bounded on every space $L^p(\mathbb{R}_+)$ $(1 \leq p < \infty)$ and can be written as

$$(W(a)\varphi)(x) = c\varphi(x) + \int_0^\infty k(x-t)\varphi(t)dt \quad (x > 0).$$

The singular integral operator S_+ on the half-line $\mathbb{R}_+$,

$$(S_+\varphi)(x) = \frac{1}{\pi i} \int_0^\infty \frac{\varphi(t)dt}{t - x} \quad (x > 0),$$

is bounded on $L^p(\mathbb{R}_+)$ if and only if $1 < p < \infty$, and it represents the simplest example of a Wiener-Hopf integral operator with piecewise continuous symbol: $S_+ = W(-\operatorname{sign} \xi)$. Note that $S_+ \in G\mathcal{B}(L^p(\mathbb{R}_+))$ if and only if $p \neq 2$.
Standard references for Fredholmness and invertibility of Wiener-Hopf integral operators are Gohberg, Krupnik [2], Gohberg, Feldman [1], Prößdorf [1], and especially Dudučava [6].

The finite section method consists in the transition to the equations

$$P_\tau W(a) P_\tau \varphi_\tau = P_\tau f, \quad \varphi_\tau \in \operatorname{Im} P_\tau,$$

where P_τ $(\tau > 0)$ is defined as follows:

$$(P_\tau\varphi)(x) = \varphi(x) \quad (0 < x < \tau), \quad (P_\tau\varphi)(x) = 0 \ (x > \tau).$$

For continuous symbols and general p as well as for piecewise continuous symbols and $p = 2$ the problem of the applicability of the finite section method was already solved in the book Gohberg, Feldman [1]. An answer to this question for piecewise continuous symbols and general p was recently obtained in Böttcher [3], [5] with the help of local methods: if $a \in (\overline{PC}_p(\mathbb{R}))_{N\times N}$ then $W(a) \in \prod_p\{P_\tau\}$ if and only if $W(a)$ and $W(\tilde{a})$ both are invertible on all spaces $L^r_N(\mathbb{R}_+)$ where $r \in [p,q]$, $1/p + 1/q = 1$ (here $\tilde{a}(\xi) := a(-\xi)$, $\xi \in \mathbb{R}$). Notice in this connection that Lemma 3.20 and its corollary (even for $N = 1$) fail to be true for Wiener-Hopf integral operators.

Two-dimensional Toeplitz operators. These operators (which more precisely are also called Toeplitz operators on the quarter-plane) are defined as follows:

$$T^{(2)}(a) : l^p(\mathbb{Z}^2_{++}) \to l^p(\mathbb{Z}^2_{++}), \quad \mathbb{Z}^2_{++} := \mathbb{Z}_+ \times \mathbb{Z}_+,$$

$$\{\varphi_{ij}\}_{i,j\in\mathbb{Z}_+} \mapsto \{\sum_{k,l\geq 0} a_{i-k,j-l}\varphi_{kl}\}_{i,j\in\mathbb{Z}_+},$$

where the so-called symbol

$$a(\xi,\eta) = \sum_{i,j\in\mathbb{Z}} a_{ij}\xi^i\eta^j, \quad (\xi,\eta) \in \Gamma^2,$$

is assumed to belong to the two-dimensional analogue of the classes introduced in 2.19. The problem of Fredholmness for such operators with continuous symbols was solved in Simonenko [5] (general p) and independently in Douglas, Howe [1] ($p = 2$): $T^{(2)}(a) \in \Phi(l^p(\mathbb{Z}^2_{++}))$ if and only if $a(\xi,\eta) \neq 0$ for $(\xi,\eta) \in \Gamma^2$ and $\mathrm{ind}_\xi a(\xi,1) = \mathrm{ind}_\eta a(1,\eta) = 0$. A two-dimensional Fredholm Toeplitz operator always has index zero, but the problem of its invertibility is of extremely difficult nature and remains open to date (compare this with the corollary in 2.8).
In this connection we mention an instructive example of Douglas, Howe [1]: if the symbol is of the form

$$a(\xi,\eta) = b(\xi\eta^{-1}) = \sum_{j\in\mathbb{Z}} b_j\xi^j\eta^{-j}, \quad (\xi,\eta) \in \Gamma^2,$$

where $b \in L^\infty(\Gamma)$, then

$$T^{(2)}(a) \in \Phi(l^2(\mathbb{Z}^2_{++})) \iff T(b) \in \Pi_2\{P_n\}$$

and

$$T^{(2)}(a) \in G\mathcal{B}(l^2(\mathbb{Z}^2_{++})) \iff T(b) \in \Pi_2\{P_n\} \text{ and}$$

<u>all</u> finite sections $T_0(b), T_1(b), T_2(b), \ldots$ of $T(b)$ are invertible.

This highlights the close relationship between Fredholmness in n dimensions and the finite section method in $n-1$ dimensions (see also Gorodeckiĭ [1], Heinig [1], Böttcher [4]).
A criterion for Fredholmness of two-dimensional Toeplitz operators on $l^2(\mathbb{Z}^2_{++})$ with symbols from the projective tensor product $\overline{PC} \hat{\otimes} \overline{PC}$ was established in Dudučava [5], and for the space $l^p(\mathbb{Z}^2_{++})$ and symbols from $\overline{PC}_p \hat{\otimes} \overline{PC}_p$ (cf. 2.42) in Böttcher [5].

The problem of the applicability of the finite section method to two-dimensional Toeplitz operators (and their various generalizations) with continuous symbols was solved in Kozak [1], [2], but see also Gorodeckiĭ [1] and Kozak, Simonenko [1]. For the case of piecewise continuous symbols results were recently obtained in Böttcher, Silbermann [4] and Böttcher [5]:

If $a(\xi,\eta)$ belongs to the projective tensor product $\overline{PC}_p(\Gamma) \hat{\otimes} \overline{PC}_p(\Gamma)$ then $T^{(2)}(a) \in \Pi_p\{P_n \otimes P_n\}$ if and only if each of the four operators

$$T^{(2)}(a),\ T^{(2)}(a_1),\ T^{(2)}(a_2),\ T^{(2)}(a_{12})$$

is invertible on $l^p(\mathbb{Z}^2_{++})$.

Here

$$a_1(\xi,\eta) = a(1/\xi,\eta), \quad a_2(\xi,\eta) = a(\xi,1/\eta),$$
$$a_{12}(\xi,\eta) = a(1/\xi, 1/\eta), \quad (\xi,\eta) \in \Gamma^2.$$

Concerning <u>two-dimensional Wiener-Hopf integral operators</u> we refer the reader to Simonenko [4] (continuous) and Dudučava [4] (PC) for Fredholmness, to Kozak [1], [2], Kozak, Simonenko [2], Widom [6] (continuous) and Böttcher [3], [5] (PC-symbols) for the finite section method.

We cite a result from Böttcher [3], [5] which highlights an interesting difference between the discrete and continuous case:

If $a(\xi,\eta)$ belongs to the projective tensor product $\overline{PC}_p(\mathbb{R}) \hat{\otimes} \overline{PC}_p(\mathbb{R})$, $1 < p < \infty$, and if $1/p + 1/q = 1$, then the finite section method with respect to the projections $P_\tau \otimes P_\tau$ is applicable to the two-dimensional Wiener-Hopf integral operator $W^{(2)}(a)$ on $L^p(\mathbb{R}^2_{++}) = L^p(\mathbb{R}_+) \hat{\otimes} L^p(\mathbb{R}_+)$ if and only if the following two conditions are satisfied:

(i) $W^{(2)}(a)$, $W^{(2)}(a_1)$, $W^{(2)}(a_2)$, $W^{(2)}(a_{12})$ are invertible on $L^p(\mathbb{R}_+) \hat{\otimes} L^p(\mathbb{R}_+)$, where

$$a_1(\xi,\eta) = a(-\xi,\eta), \quad a_2(\xi,\eta) = a(\xi,-\eta),$$
$$a_{12}(\xi,\eta) = a(-\xi,-\eta), \quad (\xi,\eta) \in \mathbb{R}^2;$$

(ii) $W^{(2)}(a) \in \Phi(L^r(\mathbb{R}_+) \hat{\otimes} L^s(\mathbb{R}_+))$ for all $r,s \in [p,q]$.

Notice the appearance of the spaces $L^r \hat{\otimes} L^s$ with mixed norm. The Fredholm theory for two-dimensional Wiener-Hopf integral operators on such spaces has been worked out in Dudučava [7].

The methods of Böttcher, Silbermann [4] and Böttcher [3], [5] are based on a bilocalization technique (cf. Pilidi [1] and Douglas, Howe [1]). While the finite section method for Toeplitz operators on $l^2(\mathbb{Z}^2_{++})$ with piecewise continuous symbols can be founded upon the local principle 1.14, we have not been able to derive the corresponding results for $l^p(\mathbb{Z}^2_{++})$ $(p \neq 2)$ by means of this principle, but were constrained to apply the (perhaps stronger) principle of Krupnik [1]. The latter principle is a generalization to the Banach algebra case of the local principle established in Douglas [2] for C*-algebras.

Unfortunately, there are no monographs dealing with two-dimensional (or higher-dimensional) Toeplitz or Wiener-Hopf operators in a detailed way. Accounts are taken in Douglas [3] and in the surveys Meister, Speck [1] and Malyšev [1], where the reader can find further results and also historical remarks.

__Effective inversion of finite Toeplitz matrices.__ There is a series of problems arising in the numerical realization of the finite section method.

For instance, it is desirable to have an estimate for the number n_o (cf. Proposition 3.3), one can ask for the speed of convergence, and there is an interest in error estimates including some knowledge about the constants occuring in them. For such question we refer the reader to Pomp [1].

Here we restrict ourselves to give some information about the effective inversion of finite Toeplitz matrices. This question is emerging not only in connection with the finite section method, but also in various other areas (momentum problems, orthogonal polynomials, stationary Gaussian processes, discrete boundary value problems), and so it has been the object of study by many people for a long time. To get an idea of the results we are concerned with here, let us quote one from Gohberg, Feldman [1], which is the culmination of several authors including Baxter, Hirschman [1], Semencul [1], and Gohberg, Semencul [1]:

Suppose the equations

$$T_n(a)x = e_o, \quad T_n(a)y = e_n \tag{35}$$

(where the notation agrees with 2.4, that is, $e_o = (\delta_{j,o})_{j=o}^n$, $e_n = (\delta_{j,n})_{j=o}^n$) have unique solutions $x = (x_j)_{j=o}^n \in \mathbb{C}^{n+1}$ and $y = (y_j)_{j=o}^n \in \mathbb{C}^{n+1}$. In addition assume $x_o \neq 0$. Then $T_n(a)$ is invertible and the inverse $T_n^{-1}(a)$ takes the form

$$T_n^{-1}(a) = \frac{1}{x_o}\left[\begin{pmatrix} x_o & & & 0 \\ x_1 & x_o & & \\ \vdots & \vdots & \ddots & \\ x_n & x_{n-1} & \cdots & x_o \end{pmatrix}\begin{pmatrix} y_n & y_{n-1} & \cdots & y_o \\ & y_n & \cdots & y_1 \\ & & \ddots & \vdots \\ 0 & & & y_n \end{pmatrix} - \right.$$

$$\left. - \begin{pmatrix} 0 & & & & 0 \\ y_o & 0 & & & \\ y_1 & y_o & 0 & & \\ \vdots & & \ddots & \ddots & \\ y_{n-1} & y_{n-2} & \cdots & y_o & 0 \end{pmatrix}\begin{pmatrix} 0 & x_n & x_{n-1} & \cdots & x_1 \\ & 0 & x_n & \cdots & x_2 \\ & & & \ddots & \vdots \\ & & & & x_n \\ 0 & & & & 0 \end{pmatrix}\right]. \tag{36}$$

Two problems which are of major interest in connection with this result can be described as follows:

(i) Find effective algorithms for the determination of the solutions of (35) and for the computation of $T_n^{-1}(a)$ on the basis of (36).

(ii) In order to remove the extra assumption $x_o \neq 0$, replace the equations (35) by others, but keep thereby the simple structure of formula (36).

With respect to (i), note that the Gaussian algorithm for the inversion of an nxn-matrix needs $O(n^3)$ computing operations. Now, assuming x and y are given, it is easy to establish a recursive procedure computing $T_n^{-1}(a)$ through (36) with only $O(n^2)$ computational steps. It is more difficult to find effective algorithms for the determination of the solutions x and y of (35) with only $O(n^2)$ operations. Under quite different hypotheses, various such algorithms were found by many authors. However, all they require that the sections $T_k(a)$ $(k = 0,1,\ldots,n-1)$ be nonsingular (notice that for a regular Toeplitz matrix $T_n(a)$ the regularity of $T_{n-1}(a)$ is equivalent to the condition $x_o \neq 0$; this is an immediate consequence of Cramer's rule).

A satisfactory result concerned with (ii) was independently found by Heinig [3] and Russakovskiĭ [1]:

Suppose the equations

$$T_n(a)x = e_o, \quad T_n(a)z = (a_{n+j-1})_{j=o}^{n} \tag{37}$$

$(a_{-1},\ldots,a_{-n}$ are the corresponding entries of $T_n(a)$ and a_{-n-1} can be chosen arbitrarily) have unique solutions $x = (x_j)_{j=o}^{n} \in \mathbb{C}^{n+1}$ and $z = (z_j)_{j=o}^{n} \in \mathbb{C}^{n+1}$. Then $T_n(a)$ is invertible and we have

$$T_n^{-1}(a) = \begin{pmatrix} z_o & & & 0 \\ z_1 & z_o & & \\ \vdots & \vdots & \ddots & \\ z_n & z_{n-1} & \cdots & z_o \end{pmatrix} \begin{pmatrix} 0 & x_n & \cdots & x_2 & x_1 \\ & & \ddots & \vdots & \vdots \\ & 0 & & 0 & x_n \\ & & & & 0 \end{pmatrix} -$$

$$- \begin{pmatrix} x_o & & & 0 \\ x_1 & x_o & & \\ \vdots & \vdots & \ddots & \\ x_n & x_{n-1} & \cdots & x_o \end{pmatrix} \begin{pmatrix} -1 & z_n & \cdots & z_2 & z_1 \\ & & \ddots & \vdots & \vdots \\ & 0 & & -1 & z_n \\ & & & & -1 \end{pmatrix} . \tag{38}$$

Note that there are no additional assumptions (of type $x_o \neq 0$) needed to ensure the validity of formula (38) (a formula which works without any extra assumptions was also given in Sachnovič [1], but its structure is of essential more intricate nature).

An effective recursive procedure for the computation of the solutions x and z of (37) was given in Rost [1]. But as well as in all earlier works, the algorithm proposed in Rost [1] is only applicable, if all sections $T_k(a)$ $(k = 0,1,\dots,n-1)$ are nonsingular. Thus the condition $x_0 \neq 0$, which to remove was the purpose of (37), (38), again appears in form of the regularity of $T_{n-1}(a)$.

Only recently Heinig [4] succeeded in establishing an inversion formula and an algorithm meeting the requirements (i) and (ii) simultaneously. The inversion formula utilized in Heinig [4] is (37), (38) and the corresponding algorithm (also working in the case where among the sections $T_0(a), T_1(a),\dots,T_{n-1}(a)$ there are singular ones) needs only $O(n^2)$ computing operations and $O(n)$ memory units. It is based on a very precise knowledge of the structure of the kernel of singular finite Toeplitz matrices and rests on a deeper analysis of some results along these lines obtained in Iochvidov [1] and Heinig [2].

All these problems will be considered in a detailed way in the book "Algebraic Methods for Toeplitz-like Operators" by G. Heinig and K. Rost, to be edited by the Akademie-Verlag as a forthcoming volume of this series. That volume will also deal with inversion formulas and effective algorithms for their analysis as well as with the structure of the kernel of more general classes of matrices (block Toeplitz and Hankel, Toeplitz-like, Vandermonde-like matrices; cf. for instance Gohberg, Heinig [1], Heinig, Rost [1], Rost [1], Kailath, Kung, Morf [1], Friedlander, Morf, Kailath, Ljung [1]). It will also be concerned with the same questions for Wiener-Hopf integral operators, singular integral operators and some others. In these investigations the so-called UV-reduction will play a central role (cf. Heinig, Rost [1], [2]).

In case $T_n(a)$ has band Toeplitz structure, an interesting procedure for the approximate solution of the equation $T_n(a)x = y$ was proposed in Dybin, Kozak [1].
Suppose $a_j = 0$ for $|j| > m$, let $n \geqq 3m-1$, and consider the equation

$$T_n(a)x = y. \qquad (39)$$

If the function

$$a(t) := \sum_{j=-m}^{m} a_j t^j \quad (t \in \Gamma) \qquad (40)$$

does not vanish on the unit circle and has winding number zero, then $a(t)$ admits a canonical Wiener-Hopf factorization $a = a_- a_+$, where $a_- \in G\overline{H^\infty}$, $a_+ \in GH^\infty$, a_- and a_+ are polynomials of degree at most m in $1/t$ and t, respectively. The mentioned algorithm can now be described as follows:

1. Find the solution x^1 of the equation $T_n(a_-)T_n(a_+)x^1 = y$ and

set $\delta^1 = y - T_n(a)x^1$.

2. Given x^{2k-1} and δ^{2k-1} $(k \geqq 1)$ find the solution x^{2k} of the equation

$$T_n(a_+)T_n(a_-)x^{2k} = \delta^{2k-1}$$

and put

$$\delta^{2k} = \delta^{2k-1} - T_n(a)x^{2k} = y - T_n(a)(x^1 + \ldots + x^{2k}).$$

3. Given x^{2k} and δ^{2k} $(k \geqq 1)$ find the solution x^{2k+1} of the equation

$$T_n(a_-)T_n(a_+)x^{2k+1} = \delta^{2k}$$

and set

$$\delta^{2k+1} = \delta^{2k} - T_n(a)x^{2k+1} = y - T_n(a)(x^1 + \ldots + x^{2k}).$$

Note that $T_n(a_-)$ and $T_n(a_+)$ are nonsingular triangular Toeplitz matrices.
The solution x of (39) is then obtained as the sum of the series

$$x = x^1 + x^2 + x^3 + \ldots + x^k + \ldots \quad . \tag{41}$$

To formulate a convergence theorem we need some preliminary facts. Set

$$r_1 = \min\{|z| \,:\, z \in \mathbb{C},\, a_+(z) = 0\}, \quad r_2 = \min\{|z| \,:\, z \in \mathbb{C},\, a_-(1/z) = 0\}.$$

By construction, $r_i > 1$ $(i = 1,2)$ and so the functions $a_+^{-1}(z)$ and $a_-^{-1}(1/z)$ are analytic in $|z| < r_1$ and $|z| < r_2$, respectively. Therefore we have a power series representation

$$a_+^{-1}(z) = \sum_{j=0}^{\infty} \alpha_j z^j \qquad (|z| < r_1),$$

$$a_-^{-1}(1/z) = \sum_{j=0}^{\infty} \beta_j z^j \quad (|z| < r_2).$$

Thus, if q_i $(i = 1,2)$ are any numbers satisfying $1/r_i < q_i < 1$, then there exist constants C_i $(i = 1,2)$ such that

$$|\alpha_j| < C_1 q_1^j, \quad |\beta_j| < C_2 q_2^j.$$

Now choose such numbers q_i $(i = 1,2)$ and set

$$\gamma_1 = \Big(\sum_{j=-m}^{m} |a_j|\Big)\Big(\sum_{j=0}^{\infty} |\beta_j|\Big) C_1 \, \frac{1-q_1^{3m-1}}{q_1^{3m-1}(1-q_1)},$$

$$\gamma_2 = \Big(\sum_{j=-m}^{m} |a_j|\Big)\Big(\sum_{j=0}^{\infty} |\alpha_j|\Big) C_2 \, \frac{1-q_2^{3m-1}}{q_2^{3m-1}(1-q_2)},$$

$$q = \max\{q_1, q_2\}, \quad \gamma = \max\{\gamma_1, \gamma_2\},$$

$$C = \left(\sum_{j=0}^{\infty} |\alpha_j|\right)\left(\sum_{j=0}^{\infty} |\beta_j|\right)\|\delta^1\| .$$

In Dybin, Kozak [1] the following result was proved:

Let $\{a_j\}_{j=-m}^{m}$ be any sequence such that the function (40) does not vanish on the unit circle and has winding number zero. Choose q as above and determine the corresponding constants γ and C. If $n \geqq 3m-1$ then

$$\|x^k\| \leqq C(\gamma q^{n+1})^{k-2} \quad \text{and} \quad \|\delta^k\| \leqq (\gamma q^{n+1})^{k-1}\|\delta^1\|.$$

Consequently, since $q < 1$, there is an $n_o \geqq 3m-1$ such that $\gamma q^{n_o+1} < 1$ and therefore, $T_n(a)$ is invertible for all $n \geqq n_o$ and the algorithm described above is applicable for such n, that is, the series (41) converges to the solution x of the equation (39).

Note that this algorithm (of course, in modified form) can also be applied if the function (40) has zeros on the unit circle (cf. Dybin, Kozak [1], p. 84–89).

CHAPTER 4

SOME BANACH ALGEBRAS RELATED WITH OPERATOR IDEALS

As will be shown in chapter 6, the investigation of the asymptotic behavior of Toeplitz determinants leads to the study of Banach algebras of functions on the unit circle which are closely related with Hankel operators from the Schatten- von Neumann classes $\mathfrak{C}_p$. Note that these algebras, in general, contain discontinuous functions. The main concern of the present chapter is the description of such algebras and the problem of invertibility in them. The next chapter will be devoted to the question of Wiener-Hopf factorization in such algebras.

BANACH ALGEBRAS GENERATED BY IDEALS

4.1. Let $\mathfrak{A}$ be a complex Banach algebra with identity e and norm $\|\cdot\|$. We assume that $\|ab\| \leqq \|a\|\|b\|$ for every $a,b \in \mathfrak{A}$. Furthermore, suppose $\mathfrak{J} \subset \mathfrak{A}$ is a two-sided ideal, which under a norm $\|\cdot\|_{\mathfrak{J}}$ becomes a Banach space, and in addition suppose the following relations hold:

(a) $\|x\| \leqq \|x\|_{\mathfrak{J}} \qquad (x \in \mathfrak{J})$

(b) $\max\{\|ax\|_{\mathfrak{J}}, \|xa\|_{\mathfrak{J}}\} \leqq \|x\|_{\mathfrak{J}} \|a\| \quad (x \in \mathfrak{J}, a \in \mathfrak{A})$.

Now let $\mathfrak{B} \subset \mathfrak{A}$ be a subalgebra and let $p_1 \in \mathfrak{A}$ be an element satisfying $p_1^2 = p_1$. Put $p_2 = e - p_1$. Then obviously $p_2^2 = p_2$.

Finally, define

$$\mathfrak{B}_0 = \{a \in \mathfrak{B} : p_1 a p_2 \in \mathfrak{J}\}.$$

Because of $p_1 e p_2 = p_1 p_2 = 0$ we have $e \in \mathfrak{B}_0$.

4.2. Lemma. The set $\mathfrak{B}_0$ is a subalgebra of $\mathfrak{B}$.

Proof.

$$p_1 ab p_2 = p_1 a(p_1 + p_2) b p_2$$

$$= p_1 a p_1 \cdot p_1 b p_2 + p_1 a p_2 \cdot p_2 b p_2. \quad \Delta$$

Note that 4.1(a),(b) were not needed in proving this lemma.

4.3. Theorem. If $\mathfrak{B}$ is a closed subalgebra of $\mathfrak{A}$, then $\mathfrak{B}_0$ is a Banach algebra under the norm

$$\|a\|_0 = (\|p_1 a p_1\|^2 + \|p_2 a p_1\|^2 + \|p_2 a p_2\|^2 + \|p_1 a p_2\|_{\mathfrak{J}}^2)^{1/2}.$$

Proof. In view of 4.1(a) for $a \in \mathfrak{B}_0$

$$\|a\| \leqq 2\|a\|_0. \qquad (1)$$

Obviously,

$$\|p_1 a p_2\|_{\mathfrak{z}} \leqq \|a\|_0. \tag{2}$$

Let now $\{a_n\} \subset \mathcal{L}_0$ be a Cauchy sequence, that is, $\|a_n - a_m\| \to 0$ as $n,m \to \infty$. Because of (1) and since $\mathcal{L}$ is supposed to be closed, there is an $a \in \mathcal{L}$ such that

$$\|a - a_n\| \to 0 \quad (n \to \infty),$$

which implies that for $j,k = 1,2$

$$\|p_j a p_k - p_j a_n p_k\| \to 0 \quad (n \to \infty). \tag{3}$$

On account of (2)

$$\|p_1 a_n p_2 - p_1 a_m p_2\|_{\mathfrak{z}} \to 0 \quad (n,m \to \infty),$$

thus there is an element $c \in \mathfrak{z}$ such that

$$\|c - p_1 a_n p_2\|_{\mathfrak{z}} \to 0 \quad (n \to \infty). \tag{4}$$

In view of 4.1(a) this gives

$$\|c - p_1 a_n p_2\| \to 0 \quad (n \to \infty). \tag{5}$$

Now from (3) $(j = 1,\ k = 2)$ and (5) it follows that $c = p_1 a_2 p_2$, i.e. $a \in \mathcal{L}_0$, and (3) and (5) imply $\|a - a_n\|_0 \to 0$ as $n \to \infty$. Thus $\mathcal{L}_0$ is closed.
Finally, let us prove that $\|ab\|_0 \leqq \|a\|_0\|b\|_0$ for every $a,b \in \mathcal{L}_0$. To that end put $c = ab$ and

$$a_{jk} = p_j a p_k, \quad b_{jk} = p_j b p_k, \quad c_{jk} = p_j c p_k \quad (j,k = 1,2).$$

Since $c_{jk} = a_{j1}b_{1k} + a_{j2}b_{2k}$, we get from 4.1(b)

$$\|c_{11}\| \leqq \|a_{11}\|\|b_{11}\| + \|a_{12}\|_{\mathfrak{z}}\|b_{21}\|,$$

$$\|c_{12}\| \leqq \|a_{11}\|\|b_{12}\|_{\mathfrak{z}} + \|a_{12}\|_{\mathfrak{z}}\|b_{22}\|,$$

hence

$$\|c_{11}\|^2 + \|c_{12}\|^2 \leqq$$

$$\leqq (\|a_{11}\|^2 + \|a_{12}\|_{\mathfrak{z}}^2)(\|b_{11}\|^2 + \|b_{21}\|^2 + \|b_{22}\|^2 + \|b_{12}\|_{\mathfrak{z}}^2).$$

Equally,

$$\|c_{21}\|^2 + \|c_{22}\|^2 \leqq$$

$$\leqq (\|a_{21}\|^2 + \|a_{22}\|^2)(\|b_{11}\|^2 + \|b_{21}\|^2 + \|b_{22}\|^2 + \|b_{12}^2\|_{\mathfrak{z}}^2)$$

and adding these two inequalities we get $\|c\|_0 \leqq \|a\|_0\|b\|_0$. Δ

Remark. The proof of the preceding theorem shows that the convergence of a sequence $\{a_n\} \subset \mathfrak{L}_0$ in the norm of $\mathfrak{L}_0$ is equivalent to its convergence in the norm of $\mathfrak{L}$ (i.e. of $\mathfrak{A}$) and to the convergence of $\{p_1 a_n p_2\}$ in the norm $\|\cdot\|_{\mathfrak{J}}$. Thus the norm $\|\cdot\|_0$ is equivalent to the norm

$$\|a\|_0' = \|a\| + \|p_1 a p_2\|_{\mathfrak{J}} \qquad (a \in \mathfrak{L}_0).$$

Thereby, for $\|p_1\| = \|p_2\| = 1$, $\|ab\|_0' \leq \|a\|_0' \|b\|_0'$ $(a,b \in \mathfrak{L}_0)$.

Now let X be a Banach space, let $\mathfrak{J} \subset \mathfrak{C}_\infty(X)$ be a two-sided ideal in $\mathfrak{B}(X)$ and let $P_1, P_2 \in \mathfrak{B}(X)$ be two complementary projections, i.e. $P_1 + P_2 = I$. According to Lemma 4.2, for every subalgebra $\mathfrak{L} \subset \mathfrak{B}(X)$ the set $\mathfrak{L}_0 = \{A \in \mathfrak{L} : P_1 A P_2 \in \mathfrak{J}\}$ forms a subalgebra of $\mathfrak{L}$.

4.4. Theorem. Assume $\mathfrak{J} \neq \{0\}$. Suppose $A \in \mathfrak{L}_0$ has an inverse $B = A^{-1}$ in $\mathfrak{L}$. If

$$P_1 A P_1 | \operatorname{Im} P_1 \in \Phi(\operatorname{Im} P_1),$$

then $B \in \mathfrak{L}_0$. For $\mathfrak{J} = \{0\}$ the conclusion remains valid, if

$$P_1 A P_1 | \operatorname{Im} P_1 \in G\mathfrak{B}(\operatorname{Im} P_1).$$

Proof. Let $\mathfrak{J} \neq \{0\}$. Put $A_1 = P_1 A P_1 + P_2 A P_2$. From $A - A_1 - P_2 A P_1 = P_1 A P_2 \in \mathfrak{C}_\infty(X)$ it follows that $A_1 + P_2 A P_1$ is Fredholm and has index zero. Let $C \in \mathfrak{B}(\operatorname{Im} P_1)$ be a regularizer of $D := P_1 A P_1 | \operatorname{Im} P_1$, that is, $CD = P_1 + P_1 T P_1$, $T \in \mathfrak{C}_\infty(\operatorname{Im} P_1)$. Then

$$(I + P_2 A P_1 C P_1)(P_1 A P_1 + P_2 A P_2) = A_1 + P_2 A P_1 + T_1,$$

where $T_1 \in \mathfrak{C}_\infty(X)$. But the operator on the right hand side is Fredholm and has index zero, and since $I + P_2 A P_1 C P_1$ is invertible (the inverse being $I - P_2 A P_1 C P_1$), the operator $P_1 A P_1 + P_2 A P_2 = A_1$ must be Fredholm with index zero. Consequently, there exist closed subspaces $N_1, N_2 \subset X$ such that

$$X = \operatorname{Ker} A_1 \oplus N_1, \quad X = \operatorname{Im} A_1 \oplus N_2$$

and thereby $\dim \operatorname{Ker} A_1 = \dim N_2$. Denote by $\hat{A}_1$ the restriction of A_1 to N_1. Then $\hat{A}_1 : N_1 \to \operatorname{Im} A$ is continuous, one-to-one, and onto, hence a continuous inverse $\hat{A}_1^{-1}$ exists. Define $B_1 \in \mathfrak{B}(X)$ to be the operator coinciding with $\hat{A}_1^{-1}$ on $\operatorname{Im} A_1$ and being zero on N_2. Denoting by $Q \in \mathfrak{C}_0(X)$ the projection onto $\operatorname{Ker} A_1$ parallel to N_1 we have for every $x \in X$,

$$B_1 A_1 x = B_1 A_1 (x - Qx) = B_1 \hat{A}_1 (x - Qx) = x - Qx,$$

that is, $B_1 A_1 = I - Q$.

Set

$$A_{jk} = P_j A P_k, \quad B_{jk} = P_j B P_k \quad (j,k = 1,2).$$

From $AB = I$ we get $A_{11}B_{12} + A_{12}B_{22} = 0$, and since $A_1 = A_{11} + A_{22}$, $A_{22}B_{12} = 0$, we obtain $A_1B_{12} + A_{12}B_{22} = 0$. On multiplying this from the left with B_1 and taking into consideration that $B_1A_1 = I - Q$, we arrive at the equality

$$B_{12} = QB_{12} - B_1A_{12}B_{22}. \tag{6}$$

From $A_{12} \in \mathfrak{J}$ we get $B_1A_{12}B_{22} \in \mathfrak{J}$. Since every two-sided non-trivial ideal in $\mathcal{B}(X)$ contains all finite rank operators, we get $QB_{12} \in \mathfrak{J}$. Then (6) implies that $B_{12} \in \mathfrak{J}$, which gives the assertion.
Now suppose $\mathfrak{J} = \{0\}$ but $P_1AP_1 | \operatorname{Im} P_1 \in G\mathcal{B}(\operatorname{Im} P_1)$. The arguments applied above yield the invertibility of A_1 and therefore $Q = 0$. From (6) we get at once that $B_{12} \in \mathfrak{J}$, which completes the proof for this case. Δ

As a by-product of the previous theorems we obtain some facts, which we have partially formulated and already proved in chapter 2.

<u>4.5. Corollary.</u> For $a \in C + H^\infty$ the Toeplitz operator $T(a)$ is Fredholm on H^2 if and only if $a \in G(C + H^\infty)$.

<u>Proof.</u> Put

$$\mathfrak{L} = \{M(a) : a \in L^\infty\} \subset \mathcal{B}(L^2)$$

and with P and Q being the operators introduced in 1.17 define

$$\mathfrak{L}_0 = \{M(a) \in \mathfrak{L} : QM(a)P \in \mathfrak{C}_\infty(L^2)\}.$$

By Theorem 2.16

$$C + H^\infty = \{a \in L^\infty : M(a) \in \mathfrak{L}_0\}.$$

Now, if $T(a) \in \Phi(H^2)$ then $a \in GL^\infty$ by Theorem 2.10, hence $M^{-1}(a) = M(a^{-1}) \in \mathcal{B}(L^2)$. Application of Theorem 4.4 gives $a^{-1} \in C + H^\infty$.

The converse is obvious. Δ

<u>Remark.</u> The argument used in this proof applies for operator-valued symbols as well.
Given a Banach space X and an X-valued summable function f on Γ define the Fourier coefficients $f_n \in X$ by

$$f_n = \frac{1}{2\pi} \int_0^{2\pi} f(e^{i\vartheta})\, e^{-in\vartheta} d\vartheta.$$

If E is a complex separable Hilbert space, then denote by $L^2(E)$

(resp. $L^\infty(\mathcal{B}(E))$) the space of weakly measurable E-valued (resp. $\mathcal{B}(E)$-valued) functions on Γ having square-summable (resp. essentially bounded) norm. Put

$$H^2(E) = \{f \in L^2(E) : f_n = 0 \text{ for } n < 0\},$$

$$H^\infty(\mathcal{B}(E)) = \{f \in L^\infty(\mathcal{B}(E)) : f_n = 0 \text{ for } n < 0\},$$

let P be the projection of $L^2(E)$ onto $H^2(E)$ defined by $(P\varphi)(t) = \sum_{n\geq 0} \varphi_n t^n$ ($|t| = 1$), and set $Q = I - P$. For $a \in L^\infty(\mathcal{B}(E))$ define the multiplication and Toeplitz operators $M(a)$ and $T(a)$ by

$$M(a) : L^2(E) \to L^2(E), \quad \varphi \mapsto a\varphi,$$

$$T(a) : H^2(E) \to H^2(E), \quad \varphi \mapsto P(a\varphi).$$

The following interesting result was obtained in Page [1]:

$$QM(a)P \in \mathfrak{C}_\infty(L^2(E)) \iff a \in C(\mathfrak{C}_\infty(E)) + H^\infty(\mathcal{B}(E)), \qquad (7)$$

where $C(\mathfrak{C}_\infty(E))$ is the space of continuous $\mathfrak{C}_\infty(E)$-valued functions on Γ. Thus $C(\mathfrak{C}_\infty(E)) + H^\infty(\mathcal{B}(E))$ is a closed subalgebra of $L^\infty(\mathcal{B}(E))$. Now we are in position to prove the following fact.

<u>Theorem.</u> If $a \in C(\mathfrak{C}_\infty(E)) + H^\infty(\mathcal{B}(E))$ then

$$T(a) \in \Phi(H^2(E)) \iff a \in G[C(\mathfrak{C}_\infty(E)) + H^\infty(\mathcal{B}(E))].$$

<u>Proof.</u> Define

$$\mathcal{L} = \{M(a) : a \in L^\infty(\mathcal{B}(E))\} \subset \mathcal{B}(L^2(E)),$$

$$\mathcal{L}_0 = \{M(a) \in \mathcal{L} : QM(a)P \in \mathfrak{C}_\infty(L^2(E))\}.$$

From (7) we get

$$C(\mathfrak{C}_\infty(E)) + H^\infty(\mathcal{B}(E)) = \{a \in L^\infty(\mathcal{B}(E)) : M(a) \in \mathcal{L}_0\}.$$

In the same fashion as in the proof of Theorem 2.10 one can show that $a \in GL^\infty(\mathcal{B}(E))$ if $T(a) \in \Phi(H^2(E))$ (define U by $(U\varphi)(t) = t\varphi(t)$). Thus, if $T(a)$ is Fredholm on $H^2(E)$ then $M(a)$ is invertible on $L^2(E)$ and Theorem 4.4 yields $a^{-1} \in C(\mathfrak{C}_\infty(E)) + H^\infty(\mathcal{B}(E))$.
The reverse conclusion is in view of (7) an obvious consequence of the vectorial analogue of Proposition 2.7. Δ

In particular, choosing $E = \mathbb{C}_N$ we obtain the following result already stated in Theorem 2.35: if $a \in (C + H^\infty)_{N\times N}$ then

$$T(a) \in \Phi(H^2_N) \iff a \in G(C + H^\infty)_{N\times N}.$$

4.6. Corollary. Denote by R the algebra of all rational functions having no poles on Γ. Then $R + H^\infty$ is an algebra and for $a \in R + H^\infty$ the operator $T(a)$ is Fredholm on H^2 if and only if $a \in G(R + H^\infty)$.

Proof. Again put

$$\mathcal{L} = \{M(a) : a \in L^\infty\} \subset \mathcal{B}(L^2)$$

but let now

$$\mathcal{L}_0 = \{M(a) \in \mathcal{L} : QM(a)P \in \mathfrak{C}_0(H^2)\}$$

(recall 1.1). By Kronecker's lemma (see 2.17(a))

$$R + H^\infty = \{a \in L^\infty : M(a) \in \mathcal{L}_0\}$$

and, consequently, $R + H^\infty$ is an algebra. If $T(a) \in \Phi(H^2)$, then $M^{-1}(a) = M(a^{-1}) \in \mathcal{B}(L^2)$ by Theorem 2.10 and therefore $a^{-1} \in R + H^\infty$ by Theorem 4.4. The reverse conclusion is trivial. Δ

4.7. Corollary. Let $a \in H^\infty$. Then

$$T(a) \in G\mathcal{B}(H^2) \iff a \in GH^\infty.$$

Proof. Choose $\mathcal{L}$ as in 4.5 or 4.6, take $\mathfrak{J} = \{0\}$ and apply Theorem 4.4. Δ

Remark. In the same way as in 4.5, one can derive vectorial analogues of the Propositions 4.6 and 4.7.

4.8. Again let X be a Banach space, $\mathfrak{J} \subset \mathfrak{C}_\infty(X)$ a two-sided ideal in $\mathcal{B}(X)$ and $P_1, P_2 \in \mathcal{B}(X)$ two complementary projections. The ideal $\mathfrak{J}$ is required to be a Banach space under a norm $\|\cdot\|_{\mathfrak{J}}$, and this norm is supposed to have the following properties:

(a) $\|A\| \leqq \|A\|_{\mathfrak{J}} \quad (A \in \mathfrak{J})$

(b) $\max\{\|BA\|_{\mathfrak{J}}, \|AB\|_{\mathfrak{J}}\} \leqq \|B\|\|A\|_{\mathfrak{J}} \quad (B \in \mathcal{B}(X),\ A \in \mathfrak{J})$.

According to Theorem 4.3 for every closed subalgebra $\mathcal{L} \subset \mathcal{B}(X)$ the set

$$\mathcal{L}_0 = \{A \in \mathcal{L} : P_1AP_2 \in \mathfrak{J}\}$$

equipped with the norm introduced in 4.3 forms a Banach algebra.
Let $\mathfrak{I} \subset \mathfrak{C}_\infty(X)$ be another ideal in $\mathcal{B}(X)$ possessing all properties required above for $\mathfrak{J}$. Define

$$\mathcal{L}_1 = \{A \in \mathcal{L} : P_2AP_1 \in \mathfrak{I}\}.$$

Then $\mathcal{L}_* = \mathcal{L}_0 \cap \mathcal{L}_1$ forms an algebra, consisting by definition of all those $A \in \mathcal{L}$, for which $P_1AP_2 \in \mathfrak{J}$ and $P_2AP_1 \in \mathfrak{I}$.

Recalling the proof of Theorem 4.3 it is not hard to show that $\mathcal{L}_*$ under the norm

$$\|A\|_* = (\|P_1AP_2\|^2 + \|P_2AP_2\|^2 + \|P_1AP_2\|_{\mathcal{J}}^2 + \|P_2AP_1\|_{\mathcal{J}}^2)^{1/2}$$

becomes a Banach algebra.

4.9. Theorem. If for $A \in \mathcal{L}_*$ there exists the inverse $B = A^{-1} \in \mathcal{L}$, then $B \in \mathcal{L}_*$.

Proof. Put $A_1 = P_1AP_1 + P_2AP_2$. Then A_1 is Fredholm and, consequently, so are $P_1AP_1|\mathrm{Im}P_1$ and $P_2AP_2|\mathrm{Im}P_2$. To complete the proof it remains to apply Theorem 4.4. Δ

BANACH ALGEBRAS GENERATED BY HANKEL OPERATORS

4.10. The algebras $K^{1/p,1/q}_{p,q}$. For $1 \leqq p \leqq \infty$ and $1 \leqq q \leqq \infty$ and with P and Q being the projections introduced in 1.17 define

$$K^{1/p,o}_{p,o} = \{a \in L^\infty: QM(a)P \in \mathfrak{C}_p(L^2)\}, \tag{8}$$

$$K^{o,1/q}_{o,q} = \{a \in L^\infty: PM(a)Q \in \mathfrak{C}_q(L^2)\}, \tag{9}$$

$$K^{1/p,1/q}_{p,q} = K^{1/p,o}_{p,o} \cap K^{o,1/q}_{o,q} . \tag{10}$$

Theorem 2.16 implies

$$K^{1/\infty,o}_{\infty,o} = C + H^\infty , \quad K^{o,1/\infty}_{o,\infty} = C + \overline{H^\infty},$$

and since $\mathfrak{C}_p \subset \mathfrak{C}_\infty$, we get at once

$$K^{1/p,o}_{p,o} \subset C + H^\infty , \quad K^{o,1/q}_{o,q} \subset C + \overline{H^\infty}.$$

Recalling Peller's characterization of the Hankel operators in $\mathfrak{C}_r(L^2)$ (see 2.17(b)) we obtain for $1 \leqq p < \infty$ and $1 \leqq q < \infty$

$$K^{1/p,o}_{p,o} = \{a \in L^\infty : Qa \in B^{1/p}_p\},$$

$$K^{o,1/q}_{o,q} = \{a \in L^\infty : Pa \in B^{1/q}_q\}.$$

From the definition and from Lemma 4.2 it follows immediately that the sets (8), (9), (10) are actually algebras. Moreover, from Theorem 4.3 and the remark in 4.3 it becomes clear that these algebras are Banach algebras under the following norms:

$$\|a\| = \|a\|_{L^\infty} + \|Qa\|_{B^{1/p}_p} \qquad (a \in K^{1/p,o}_{p,o}),$$

$$\|a\| = \|a\|_{L^\infty} + \|Pa\|_{B_q^{1/q}} \qquad (a \in K_{o,q}^{o,1/q}),$$

$$\|a\| = \|a\|_{L^\infty} + \|Qa\|_{B_p^{1/p}} + \|Pa\|_{B_q^{1/q}} \qquad (a \in K_{p,q}^{1/p,1/q}).$$

4.11. Theorem. A function in $K_{p,o}^{1/p,o}$ (resp. $K_{o,q}^{o,1/q}$) is invertible in $K_{p,o}^{1/p,o}$ (resp. $K_{o,q}^{o,1/q}$) if and only if it is invertible in $C + H^\infty$ (resp. $C + \overline{H^\infty}$). For $a \in K_{p,q}^{1/p,1/q}$ we have $a \in GK_{p,q}^{1/p,1/q} \iff$ $\iff a \in G(C + H^\infty) \iff a \in G(C + \overline{H^\infty}) \iff a \in GL^\infty$.

Proof. The necessity portions are obvious. To prove the sufficiency put

$$\mathcal{L} = \{M(\varphi) : \varphi \in L^\infty\} \subset \mathcal{B}(L^2),$$

$$\mathcal{L}_0 = \{M(\varphi) \in \mathcal{L} : QM(\varphi)P \in \mathfrak{C}_p(L^2)\}.$$

Now, if $a \in K_{p,o}^{1/p,o}$ is invertible in $C + H^\infty$ then, obviously, $M(a) \in G\mathcal{L}$, $M(a) \in \mathcal{L}_0$, and, by virtue of Corollary 4.5, the operator $PM(a)P|\operatorname{Im} P$ is Fredholm. Thus Theorem 4.4 applies and it results $M^{-1}(a) = M(a^{-1}) \in \mathcal{L}_0$, that is, $a^{-1} \in K_{p,o}^{1/p,o}$.

Suppose now $a \in K_{p,q}^{1/p,1/q}$ is invertible in L^∞. Let $\mathcal{L}$ and $\mathcal{L}_0$ be as above and set

$$\mathcal{L}_1 = \{M(\varphi) \in \mathcal{L} : PM(\varphi)Q \in \mathfrak{C}_q(L^2)\},$$
$$\mathcal{L}_* = \mathcal{L}_0 \cap \mathcal{L}_1.$$

Then $M(a) \in G\mathcal{L}$ and $M(a) \in \mathcal{L}_*$, consequently, $M^{-1}(a) = M(a^{-1}) \in \mathcal{L}_*$, i.e. $a^{-1} \in K_{p,q}^{1/p,1/q}$, by Theorem 4.9. Δ

4.12. Quasicontinuous functions. The algebra

$$K_{\infty,\infty}^{1/\infty,1/\infty} = (C + H^\infty) \cap (C + \overline{H^\infty})$$

is usually called the algebra of quasicontinuous functions and denoted by QC. Note that $C \subset QC$ but $C \neq QC$. The following well-known result is an immediate consequence of Theorem 4.11 (see also 2.41(e), (f)): if $a \in QC$ then

$$a \in G(QC) \iff a \in GL^\infty \iff T(a) \in \Phi(H^2).$$

4.13. Multiplication and Toeplitz operators on weighted l^p-spaces.

For $1 < p < \infty$ and $\alpha,\beta \in \mathbb{R}$ let $l_p^{\alpha,\beta}$ be the Banach space of all sequences $\varphi = \{\varphi_j\}_{j\in\mathbb{Z}}$ satisfying

$$\|\varphi\|^p = \sum_{j=-\infty}^{-1} |\varphi_j|^p(|j|+1)^{p\alpha} + \sum_{j=0}^{\infty} |\varphi_j|^p(j+1)^{p\beta} < \infty. \tag{11}$$

Note that $l_p^{0,0}$ is just the space $\tilde{l}^p$ introduced in 2.1. By $M_{p,\gamma}$ we denote the algebra of all functions $a \in L^1$ such that $M(a) \in \mathcal{B}(l_p^{\gamma,\gamma})$. For a detailed investigation of the classes $M_{p,\gamma}$ we refer the reader to Verbickiĭ [1]. Here should be noted only the following facts (let $1/p + 1/q = 1$):

(a) $M_{p,\gamma} = M_{q,-\gamma} \subset M_{p,0} = M_{q,0}$ for all $1 < p < \infty$ and $\gamma \in \mathbb{R}$.

(b) If $\gamma > 1/q$ then

$$M_{p,\gamma} = \{a \in L^1 : \sum_{k\in\mathbb{Z}} |a_k|^p(|k| + 1)^{p\gamma} < \infty\}$$

and for $\gamma < -1/p$

$$M_{p,\gamma} = \{a \in L^1 : \sum_{k\in\mathbb{Z}} |a_k|^q(|k| + 1)^{-\gamma q} < \infty\}.$$

(c) If $-1/p < \gamma < 1/q$ then $M_{p,\gamma}$ contains all functions $a \in L^\infty$ of bounded variation.

Later we shall need the following property of $M_{2,\gamma}$ (see Verbickiĭ [1], Theorem 6',II):

(d) If $|\gamma| < \mu < 1$ then $C^\mu \subset M_{2,\gamma}$.

Now put

$$l_p^\gamma = \{\varphi \in l_p^{\gamma,\gamma} : \varphi_k = 0 \quad \text{for} \quad k < 0\}. \tag{12}$$

Then $T(a) \in \mathcal{B}(l_p^\gamma)$ for $a \in M_{p,\gamma}$. If $\gamma = 0$ then $a \in M_{p,0}$ is a consequence of the boundedness of $T(a)$ on $l_p^0 = l^p$ (see Theorem 2.5). For $\gamma \neq 0$, however, there are $a \notin M_{p,\gamma}$ such that $T(a) \in \mathcal{B}(l_p^\gamma)$.

Finally, Toeplitz operators may be considered on other spaces as well. So, under appropriate conditions, $T(a)$ defined by $T(a)\varphi = P(a\varphi)$ will be bounded on $(B_p^\alpha)_A = PB_p^\alpha$ (recall 1.20).

4.14. Hankel operators on weighted l^2-spaces. For $\alpha \geq 0$, $\beta \geq 0$ let $W_2^{\alpha,\beta}$ be the class of all functions $a \in L^2$ having the property that that the sequence of Fourier coefficients belongs to $l_2^{\alpha,\beta}$. Define

$W_2^{-\alpha,-\beta} = (W_2^{\alpha,\beta})^*$, $W_2^{\alpha,-\beta} = W_2^{\alpha,0} + (W_2^{0,\beta})^*$, $W_2^{-\alpha,\beta} = (W_2^{\alpha,0})^* + W_2^{0,\beta}$.

The elements of $W_2^{\alpha,\beta}$, where $\alpha < 0$ or $\beta < 0$, may be regarded as distributions on Γ. Henceforth we shall not make a distinction between functions and the sequence of their Fourier coefficients, that is, we shall identify $W_2^{\alpha,\beta}$ with $l_2^{\alpha,\beta}$ and prefer $l_2^{\alpha,\beta}$ as notation for this class.

In Peller [3] the following facts are proved:

(a) Let $1 \leqq p < \infty$, $\alpha > -\frac{1}{2}$, $\alpha > -\frac{1}{p}$, $\beta > -\frac{1}{2}$, $\beta > -\frac{1}{p}$.

Then

$$QM(a)P \in \mathfrak{C}_p(l_2^{\alpha,-\beta}) \iff Qa \in B_p^{1/p+\alpha+\beta}.$$

(b) Let $1 \leqq p < \infty$, $\alpha < \frac{1}{2}$, $\alpha < \frac{1}{q}$, $\beta < \frac{1}{2}$, $\beta < \frac{1}{q}$.

Then

$$PM(a)Q \in \mathfrak{C}_q(l_2^{\alpha,-\beta}) \iff Pa \in B_q^{1/q-\alpha-\beta}.$$

In particular,

(c) For $1 \leqq p < \infty$, $\gamma > -\frac{1}{2}$, $\gamma > -\frac{1}{p}$,

$$QM(a)P \in \mathfrak{C}_p(l_2^{\gamma,0}) \iff Qa \in B_p^{1/p+\gamma}.$$

(d) For $1 \leqq p < \infty$, $\gamma < \frac{1}{2}$, $\gamma < \frac{1}{q}$,

$$PM(a)Q \in \mathfrak{C}_q(l_2^{\gamma,0}) \iff Pa \in B_q^{1/q-\gamma}.$$

Furthermore (see also Peller [3]),

(e) Let $\alpha > 0$, $\beta > 0$. Then

$$QM(a)P \in \mathfrak{B}(l_2^{\alpha,-\beta}) \iff Qa \in C^{\alpha+\beta};$$

$$QM(a)P \in \mathfrak{C}_\infty(l_2^{\alpha,-\beta}) \iff Qa \in c^{\alpha+\beta}.$$

Note that in view of $B_2^\delta = l_2^{\delta,\delta}$ the assertions (a) - (d) for $p = q = 2$ are easy to verify by a direct computation based on 1.5.

<u>4.15. The classes $K_{p,q}^{\alpha,\beta}$.</u> Suppose

$$p > 1, \quad q > 1, \quad \frac{1}{p} + \frac{1}{q} = 1, \quad \alpha > 0, \ \beta > 0, \ \alpha + \beta = 1, \tag{13}$$

and define

$$K_{p,q}^{\alpha,\beta} = \{a \in L^\infty : Qa \in B_p^\alpha, \ Pa \in B_q^\beta\}. \tag{14}$$

In the case where besides (13) the condition

$$-\frac{1}{2} < \alpha - \frac{1}{p} < \frac{1}{2} \quad \text{(equivalently, } -\frac{1}{2} < \beta - \frac{1}{q} < \frac{1}{2}) \tag{15}$$

holds, we shall give another characterization of the class (14). Because of (13) there is a real number γ such that

$$\alpha = \frac{1}{p} + \gamma, \quad \beta = \frac{1}{q} - \gamma, \quad -\frac{1}{p} < \gamma < \frac{1}{q},$$

and due to (15) we have additionally $-\frac{1}{2} < \gamma < \frac{1}{2}$. Now 4.13(c), (d) give

$$K_{p,q}^{\alpha,\beta} = K_{p,q}^{\frac{1}{p}+\gamma, \frac{1}{q}-\gamma}$$

$$= \{a \in L^\infty: QM(a)P \in \mathfrak{C}_p(l_2^{\gamma,0}),\ PM(a)Q \in \mathfrak{C}_q(l_2^{\gamma,0})\}.$$

The property which will make this class so interesting for us, consists in the fact that under the hypotheses (13), (15) for $a,b \in K_{p,q}^{\alpha,\beta}$ the operator $H(a)H(\tilde{b})$ belongs to $\mathfrak{C}_1(H^2)$.
This is a consequence of the identity

$$H(a)H(\tilde{b}) = PM(a)Q \cdot QM(b)P$$

and of $PM(a)Q \in \mathfrak{C}_q(l_2^{\gamma,0})$, $QM(b)P \in \mathfrak{C}_p(l_2^{\gamma,0})$.
If $p = q = 2$ then $K_{2,2}^{\alpha,\beta} = L^\infty \cap l_2^{\alpha,\beta}$, and in case $\alpha = 1/p$, $\beta = 1/q$ we obtain the algebras defined by 4.10(10). It turns out that $K_{p,q}^{\alpha,\beta}$ is an algebra if only (13), (15) are satisfied. But the proof of this fact is essentially more difficult than in the case $\alpha = 1/p$, $\beta = 1/q$. Unfortunately, it is also not so easy as for that case to show that an element in $K_{p,q}^{\alpha,\beta}$ $(\alpha \neq 1/p,\ \beta \neq 1/q)$ is invertible in $K_{p,q}^{\alpha,\beta}$ if and only if it is invertible in $C + H^\infty$ $(\alpha > 1/p)$ respectively $C + \overline{H^\infty}$ $(\alpha < 1/p)$. The remainder of this chapter is devoted to the proofs of those two facts.

<u>4.16. Lemma.</u> If $1 < p < \infty$ and $1/p < \alpha < 1$ then $B_p^\alpha \subset C$.

<u>Proof.</u> This is well-known (cf., for instance, Triebel [1], relations (10), p. 71 and (12), p. 103). Δ

<u>Corollary 1.</u> Under the condition 4.15(13) the set $K_{p,q}^{\alpha,\beta}$ is contained in $C + H^\infty$ for $\alpha \geq 1/p$, and in $C + \overline{H^\infty}$ for $\beta \geq 1/q$.

<u>Proof.</u> For $\alpha = 1/p$ see 4.10 and for $\alpha > 1/p$ apply the preceding lemma. Δ

<u>Corollary 2.</u> Under the condition 4.15(13) and for $\alpha \neq 1/p$ (or, equivalently, $\beta \neq 1/q$) the projections P and Q are bounded in $K_{p,q}^{\alpha,\beta}$.

<u>Proof.</u> Apply the above lemma and the closed graph theorem. Δ

4.17. Lemma. Suppose $1 < p < \infty$ and $0 < \alpha < 1$. Then $L^\infty \cap B_p^\alpha$ is an algebra, and if $\alpha > 1/p$ then B_p^α itself forms an algebra.

Proof. According to Lemma 4.16 we have $B_p^\alpha = L^\infty \cap B_p^\alpha$ for $\alpha > 1/p$. Hence it sufficies to verify the first assertion. Thus let $a \in L^\infty \cap B_p^\alpha$. Then, obviously, $a^2 \in L^\infty$ and in view of $\|a^2\|_{L^p} \leq \|a\|_{L^\infty}\|a\|_{L^p}$, we find that $a^2 \in L^p$. Furthermore,

$$\|a^2(e^{i(x+t)}) - a^2(e^{ix})\|_{L^p}^p$$

$$= \|[a(e^{i(x+t)}) + a(e^{ix})][a(e^{i(x+t)}) - a(e^{ix})]\|_{L^p}^p$$

$$\leq 2\|a\|_{L^\infty}^p\|a(e^{i(x+t)}) - a(e^{ix})\|_{L^p}^p$$

and it remains to recall how B_p^α was defined in 1.20 to conclude that $a^2 \in B_p^\alpha$. Now the assertion follows from

$$ab = \frac{1}{2}[(a + b)^2 - a^2 - b^2]. \ \Delta$$

4.18. Lemma. Let $a_- \in \overline{H^\infty}$. Then the Toeplitz operator $T(a_-)$ is bounded on $(B_p^\alpha)_A = PB_p^\alpha$ for $1 < p \leq \infty$ and $0 < \alpha < 1$.

Proof. The adjoint of $(B_p^\alpha)_A$ is $(B_q^{-\alpha})_A$, $1/p + 1/q = 1$. Thus the adjoint $T(\overline{a}_-)$ of $T(a_-)$ can be identified with the multiplication operator $M(\overline{a}_-)$ on $(B_q^{-\alpha})_A$.
If $q = 1$ $(p = \infty)$ then according to Peller, Hruščev [1], II, p.20,

$$\varphi \in (B_1^{-\alpha})_A \Longleftrightarrow \iint_{\mathbb{D}} (1 - |z|)^{\alpha-1}|\hat{\varphi}(z)|\,dxdy < \infty$$

and the boundedness of $M(\overline{a}_-)$ becomes obvious.
Thus let $1 < q < \infty$ $(1 < p < \infty)$. Then by 1.20

$$\varphi \in (B_q^{-\alpha})_A \Longleftrightarrow I_1\varphi \in (B_q^{1-\alpha})_A$$

$$\Longleftrightarrow \int_0^1 (1-r)^{1q-(1-\alpha)q-1} \|(I_1\varphi)_r'\|_{L^p}^p dr < \infty$$

$$\Longleftrightarrow \int_0^1 (1-r)^{\alpha q-1} \|\varphi_r\|_{L^p}^p dr < \infty$$

and the boundedness of $M(\overline{a}_-)$ follows from

$$\|(\overline{a}_-\varphi)_r\|_{L^p} = \|(\overline{a}_-)_r\varphi_r\|_{L^p} \leq \|(\overline{a}_-)_r\|_{L^\infty}\|\varphi_r\|_{L^p} \ . \ \Delta$$

4.19. Theorem. Under the conditions 4.15(13) the set $K_{p,q}^{\alpha,\beta}$ forms an algebra, which provided with the norm

$$\|a\| = \|a\|_{L^\infty} + \|Qa\|_{B_p^\alpha} + \|Pa\|_{B_q^\beta} \qquad (16)$$

becomes a Banach algebra.

Proof. For $\alpha = 1/p$, and therefore $\beta = 1/q$, this was shown in 4.10. Thus there is no loss of generality in assuming $\alpha > 1/p$ (which implies that $\beta < 1/q$). Let $a, b \in K^{\alpha,\beta}_{p,q}$ and set

$$a_+ = Pa, \quad a_- = Qa, \quad b_+ = Pb, \quad b_- = Qb.$$

By Corollary 2 in 4.16 we have $a_\pm, b_\pm \in K^{\alpha,\beta}_{p,q}$. Now

$$ab = a_+b_+ + a_+b_- + a_-b_+ + a_-b_- .$$

But a_+b_+ and a_-b_- are in $K^{\alpha,\beta}_{p,q}$ by Lemma 4.17, whereas a_+b_- and a_-b_+ belong to $K^{\alpha,\beta}_{p,q}$ due to Lemma 4.18. Thus $K^{\alpha,\beta}_{p,q}$ forms an algebra. That it is a Banach algebra under the norm (16) is now obvious. Δ

4.20. Proposition. If $a \in L^\infty$ and $0 < \gamma < 1$ then

$$M(a) \in \mathcal{B}(l_2^{\gamma/2,-\gamma/2}) \iff a \in C^\gamma + H^\infty.$$

Proof. The necessity portion follows from 4.14(e). For the sufficiency part take $a \in C^\gamma + H^\infty$ and $\varphi \in l_2^{\gamma/2,-\gamma/2}$, put $a_+ = Pa$, $a_- = Qa$, $\varphi_+ = P\varphi$, $\varphi_- = Q\varphi$ and write

$$a\varphi = P(a_+\varphi_+) + P(a_+\varphi_-) + P(a_-\varphi_+) + Q(a_+\varphi_-) + Q(a_-\varphi_+) + Q(a_-\varphi_-).$$

Now, $QM(a_-)P \in \mathcal{B}(l_2^{\gamma/2,-\gamma/2})$ by 4.14(e); $PM(a_-)P$ and $QM(a_-)Q$ are bounded on $l_2^{\gamma/2,-\gamma/2}$ due to 4.13(d) and the well-known fact that P and Q are bounded in C^γ; Lemma 4.18 implies that $PM(a_+)P = T(\overline{a}_+)^*$ and $QM(a_+)Q$ are in $\mathcal{B}(l_2^{\gamma/2,-\gamma/2})$; finally, that $PM(a_+)Q$ belongs to $\mathcal{B}(l_2^{\gamma/2,-\gamma/2})$ is obvious. Δ

Remark. The assumption "$a \in L^\infty$" is redundant since it can be shown to follow automatically from the boundedness of $M(a)$ on $l_2^{\gamma/2,-\gamma/2}$ (although this is not trivial).

4.21. Proposition. Suppose $\gamma > 0$. Then $C^\gamma + H^\infty$ and $c^\gamma + H^\infty$ are algebras. Under the norm

$$\|a\| = \|a\|_{L^\infty} + \|M(a)\|_{\mathcal{B}(l_2^{\gamma/2,-\gamma/2})} \tag{17}$$

they become Banach algebras.

Proof. For $C^\gamma + H^\infty$ this is obvious from the preceding proposition. Now put $\mathcal{A} = \mathcal{B}(l_2^{\gamma/2,-\gamma/2})$ and define

$$\mathcal{L} = \{M(a) : a \in C^\gamma + H^\infty\} \subset \mathcal{A}.$$

Then $\mathcal{L}$ is a closed subalgebra of $\mathcal{A}$ and by Theorem 4.3,

$$\mathcal{L}_0 = \{M(a) \in \mathcal{L} : QM(a)P \in \mathfrak{C}_\infty(l_2^{\gamma/2,-\gamma/2})\}$$

provided with the norm $\|\cdot\|_{\mathfrak{A}}$ becomes a Banach algebra. Because of 4.14(e)

$$\mathcal{L}_0 = \{M(a) : a \in c^\gamma + H^\infty\} \subset \mathfrak{A}\,.$$

Hence $c^\gamma + H^\infty$ is a Banach algebra under the norm (17). Δ

<u>Remark.</u> Recalling the remark in 4.20 we see that this proposition remains true if norm (17) is replaced by the norm $\|a\| = \|M(a)\|_{\mathcal{B}(l_2^{\gamma/2,-\gamma/2})}$.

<u>4.22. Lemma.</u> Suppose $p > 1$, $q > 1$, $1/p + 1/q = 1$, and $0 < \gamma < \min\{1/q, 1/2\}$. Then the set

$$K_{p,q}^{\frac{1}{p}+\gamma,\frac{1}{q}-\gamma} \cap (c^\gamma + H^\infty)$$

forms an algebra. An element of this algebra is invertible in this algebra if and only if it is invertible in $c^\gamma + H^\infty$. Provided with the norm

$$\|a\| = \|a\|_{L^\infty} + \|M(a)\|_{\mathcal{B}(l_2^{\gamma/2,-\gamma/2})} + \|Qa\|_{B_p^{1/p+\gamma}} + \|Pa\|_{B_q^{1/q-\gamma}} \qquad (18)$$

this algebra becomes a Banach algebra.

<u>Proof.</u> Due to Proposition 4.21 the set $\mathcal{L}$ defined as

$$\mathcal{L} = \{M(a) : a \in c^\gamma + H^\infty\} \subset \mathcal{B}(l_2^{\gamma/2,-\gamma/2})$$

is a closed subalgebra of $\mathcal{B}(l_2^{\gamma/2,-\gamma/2})$. Thus by 4.8 the set

$$\mathcal{L}_* = \{M(a) \in \mathcal{L} : QM(a)P \in \mathfrak{C}_p(l_2^{\gamma/2,-\gamma/2}),$$
$$PM(a)Q \in \mathfrak{C}_q(l_2^{\gamma/2,-\gamma/2})\}$$

provided with the norm (18) is a Banach algebra. Because of 4.14(a),(b)

$$\mathcal{L}_* = \{M(a) \in \mathcal{L} : Qa \in B_p^{\frac{1}{p}+\gamma},\ \ Pa \in B_q^{\frac{1}{q}-\gamma}\},$$

therefore $\mathcal{L}_*$ admits the following description:

$$\mathcal{L}_* = \{M(a) : a \in (c^\gamma + H^\infty) \cap K_{p,q}^{\frac{1}{p}+\gamma,\frac{1}{q}-\gamma}\} \subset \mathcal{B}(l_2^{\gamma/2,-\gamma/2}).$$

Thus if $M(a) \in \mathcal{L}_*$ then Theorem 4.9 gives

$$M(a) \in G\mathcal{L}_* \iff M(a) \in G\mathcal{L},$$

which is equivalent to our assertion. Δ

<u>4.23. Proposition.</u> Let $a \in c^\gamma + H^\infty$ $(\gamma > 0)$. Then

$$a \in G(c^\gamma + H^\infty) \Longleftrightarrow a \in G(C + H^\infty).$$

<u>Proof.</u> Obviously, Qc^γ and H^∞ are closed subalgebras of $c^\gamma + H^\infty$ and

$$c^\gamma + H^\infty = \text{clos}\{\overline{\chi}_n\psi : n \in \{0,1,2,\dots\},\ \psi \in H^\infty\}$$

(the norm in $c^\gamma + H^\infty$ being, of course, defined by 4.21(17)).
Now, in a similar fashion as in Douglas [2], Proposition 6.37, we shall prove that the maximal ideal space N of $c^\gamma + H^\infty$ can be identified with those elements of the maximal ideal space M of H^∞ satisfying

$$|\chi_k(m)| = 1 \quad \text{for all} \quad k \in \{0,1,2,\dots\}; \tag{19}$$

here φ denotes the Gelfand transform of φ.

If f is a multiplicative linear functional on $c^\gamma + H^\infty$, then $f|H^\infty$ is a multiplicative linear functional on H^∞, and hence $\Psi : f \mapsto f|H^\infty$ defines a continuous mapping of N into M. If f_1 and f_2 are elements in N such that $\Psi(f_1) = \Psi(f_2)$, then for $k = 0,1,2,\dots$

$$f_1(\overline{\chi}_k) = f_1(\chi_k^{-1}) = f_1(\chi_k)^{-1} = f_2(\chi_k)^{-1} = f_2(\chi_k^{-1}) = f_2(\overline{\chi}_k)$$

and thus $f_1 = f_2$. Therefore, Ψ is a homeomorphism of N into M. Moreover, since the spectral radius of χ_k and $\overline{\chi}_k$ equals 1, we have for f in N

$$|f(\chi_k)| \leqq 1 \quad \text{and} \quad 1/|f(\chi_k)| = |f(\overline{\chi}_k)| \leqq 1,$$

i.e. $|f(\chi_k)| = 1$ for $k = 0,1,2,\dots$. Consequently, the image of Ψ is contained in

$$\{m \in M : |\chi_k(m)| = 1 \quad \text{for} \quad k \in \{0,1,2,\dots\}\}.$$

Let m now be a point in M such that $|\chi_k(m)| = 1$ for $k=0,1,2,\dots$
If we define g on $\{\overline{\chi}_k\psi : k \in \{0,1,2,\dots\},\ \psi \in H^\infty\}$ such that $g(\overline{\chi}_k\psi) = \psi(m)\chi_k(m)$, then g can easily be shown to be multiplicative, and the inequality

$$|g(\overline{\chi}_k\psi)| = |\psi(m)||\chi_k(m)| = |\psi(m)| \leqq \|\psi\|_{H^\infty} =$$

$$= \|\overline{\chi}_k\psi\|_{L^\infty} \leqq \|\overline{\chi}_k\psi\|_{c^\gamma+H^\infty}$$

shows that g can be extended to a multiplicative linear functional on $c^\gamma + H^\infty$.
Obviously, $\Psi(g) = m$ and thus N is homeomorphic to that part of M which is given by (19).

According to Douglas [2], Proposition 6.37, the maximal ideal space of $C + H^\infty$ can also be identified with that part of M which is characterized by (19). This completes the proof. Δ

4.24. Theorem. Suppose $p > 1$, $q > 1$, $1/p + 1/q = 1$, $0 < \gamma < \min\{1/q, 1/2\}$, and let $a \in K_{p,q}^{1/p+\gamma, 1/q-\gamma}$. Then

$$a \in GK_{p,q}^{\frac{1}{p}+\gamma, \frac{1}{q}-\gamma} \Longleftrightarrow a \in G(C + H^\infty).$$

Proof. Put

$$A = K_{p,q}^{\frac{1}{p}+\gamma, \frac{1}{q}-\gamma}, \quad B = K_{p,q}^{\frac{1}{p}+\gamma, \frac{1}{q}-\gamma} \cap (C^\gamma + H^\infty).$$

Then $C + H^\infty \supset A \supset B$ and B is continuously and dense embedded into A. Given a commutative Banach algebra $\mathcal{A}$ we denote by $M(\mathcal{A})$ the maximal ideal space of $\mathcal{A}$, thought of as identified with the space of multiplicative linear functionals on $\mathcal{A}$. If $b \in B$ then by Lemma 4.22 and Proposition 4.23

$$b \in GB \Longleftrightarrow b \in G(C + H^\infty).$$

Therefore the spectrum of $b \in B$ in B coincides with that of b in $C + H^\infty$. Hence for each $\varphi \in M(B)$ there is a $\psi \in M(C + H^\infty)$ (depending on φ and b) such that

$$\psi(b) = \varphi(b). \tag{20}$$

Assume now there exists an $a \in A$ being invertible in $C + H^\infty$ but not in A. Thus we can find a $\varphi \in M(A)$ with $\varphi(a) = 0$. Furthermore, there is a sequence $\{b_n\} \subset B$ such that

$$\|a - b_n\|_A \to 0 \quad (n \to \infty), \tag{21}$$

consequently, $\varphi(b_n) \to 0$ as $n \to \infty$. Since $\varphi \in M(B)$, it follows from (20) that there are $\psi_n \in M(C + H^\infty)$ such that

$$\psi_n(b_n) = \varphi(b_n) \to 0 \quad (n \to \infty). \tag{22}$$

On the other hand, from (21) we get $\|a - b_n\|_{L^\infty} \to 0$ as $n \to \infty$ and since $a \in G(C + H^\infty)$, it results the invertibility of b_n in $C + H^\infty$ for all sufficiently large n. Thereby $\|b_n^{-1} - a^{-1}\|_{L^\infty} \to 0$ as $n \to \infty$ and, consequently, $\|b_n^{-1}\|_{L^\infty} \leqq 2\|a^{-1}\|_{L^\infty}$ for all n large enough.

Since every multiplicative linear functional has norm 1, we obtain

$$|\psi_n(b_n^{-1})| \leqq 2\|a^{-1}\|_{L^\infty} \tag{23}$$

for all sufficiently large n. But $\psi_n(b_n)\, \psi_n(b_n^{-1}) = 1$ and thus (22) and (23) are contradictory. This completes the proof. Δ

Putting Theorems 4.11 and 4.24 together we get the following:

Assume the conditions 4.15(13), (15) are fulfilled and let $a \in K^{\alpha,\beta}_{p,q}$. Then $a \in GK^{\alpha,\beta}_{p,q}$ is equivalent to

$$a \in G(C + H^\infty) \quad \text{if} \quad \alpha > 1/p\,;$$
$$a \in G(C + \overline{H^\infty}) \quad \text{if} \quad \alpha < 1/p\,;$$
$$a \in GL^\infty \quad \text{if} \quad \alpha = 1/p\,.$$

NOTES AND COMMENTS

4.3, 4.9. These results are due to Krein [2].

4.4. - 4.7. Theorem 4.4 is published here for the first time, although its proof makes essentially use of the ideas of Krein [2]. The Corollaries 4.5 and 4.7 are well-known (see 2.18 and 2.12), whereas Corollary 4.6 and the theorem in 4.5 as well as the method to obtain all these results seem to be new. For vectorial Hankel operators confer Peller [3] and the references cited there.

4.10, 4.11, 4.15. The algebra $K^{1/2,1/2}_{2,2}$ was introduced and studied in Krein [2]. Its generalization is due to the authors.

4.13. See Verbickiĭ [1].

4.14. See Peller [3].

4.18. Various versions of this lemma were found by several authors. The proof presented in text was suggested by Peller, Hruščev [1], II, p.20.

4.19. - 4.24. These results are due to the authors and are published here for the first time.

CHAPTER 5

FACTORIZATION IN ALGEBRAS OF ESSENTIALLY BOUNDED MATRIX FUNCTIONS

The theory of Toeplitz operators and their various generalizations is closely related to the question of Wiener-Hopf factorization. This chapter deals with this problem for some algebras of essentially bounded functions on the unit circle.

FORMULATION OF THE MAIN RESULT

5.1. Decomposing algebras. Let A be an algebra of complex-valued functions on the unit circle Γ possessing the following properties:

(a) A is a Banach algebra (under a norm $\|\cdot\|_A$), which is continuously embedded in L^∞.

(b) A contains all functions χ_k, $k \in \mathbb{Z}$ (see 2.4).

The projections P and Q defined in 1.17 are bounded on L^p for $1 < p < \infty$. Therefore $PA \subset L^p$ and $QA \subset L^p$ for all p, $1 < p < \infty$.

Definition. An algebra A is called decomposing algebra, if it has besides the properties (a) and (b) the following property:

(c) $PA \subset A$ and $QA \subset A$.

By means of the closed graph theorem it is easy to deduce from (c) that P and Q are bounded on A and that PA and QA are closed subspaces of A.

For a decomposing algebra A put $A_+ = PA$ and $\overset{\circ}{A}_- = QA$. By A_0 we denote the space of constant functions on Γ, and we define $\overset{\circ}{A}_+ = \chi_1 A_+$ and $A_- = A_0 \oplus \overset{\circ}{A}_-$. Then

$$A = \overset{\circ}{A}_- \oplus A_+ = A_- \oplus \overset{\circ}{A}_+.$$

5.2. Factorization. Given a set M we denote by M^n, M_n, $M_{n\times n}$ the sets of $1\times n$, $n\times 1$, $n\times n$ matrices with entries in M, respectively. If A is a decomposing algebra, then $(A_{\pm})_n$, $(\overset{\circ}{A}_{\pm})_n$ etc. will be written as $A_n^{\pm}$, $A_n^{\circ\pm}$ etc.

Definition. Let A be a decomposing algebra. A matrix function $a(t) \in A_{n\times n}$ is said to admit a (right) factorization in A, if

$$a(t) = a_-(t)d(t)a_+(t) \qquad (t \in \Gamma),$$

where

$$a_{\pm}(t) \in GA^{\pm}_{n\times n},$$

$$d(t) = \operatorname{diag}(t^{\varkappa_1},\dots,t^{\varkappa_n}),$$

$$\varkappa_i \in \mathbb{Z} \quad (i = 1,\dots,n), \qquad \varkappa_1 \leqq \varkappa_2 \leqq \dots \leqq \varkappa_n.$$

The integers $\varkappa_i$ are called (right) partial indices of $a(t)$.

5.3. Factorization property. Let A be a decomposing algebra. If $a(t) \in A_{n\times n}$ admits a factorization in A then obviously $a(t) \in GA_{n\times n}$.

Definition. A decomposing algebra A is said to have the factorization property, if every matrix function in $GA_{n\times n}$ admits a factorization in A.

It is the purpose of this chapter to establish necessary and sufficient conditions for a decomposing algebra to have the factorization property.

5.4. Fredholm property. Let A be a decomposing algebra. Given a matrix function $a(t) \in A_{n\times n}$ we define two operators $T_+(a)$ and $T_-(a)$, which will be important throughout our further investigations:

$$T_+(a) : A^+_n \to A^+_n, \quad f_+ \mapsto P(af_+) ;$$

$$T_-(a) : \overset{o}{A}{}^n_- \to \overset{o}{A}{}^n_-, \quad f_- \mapsto Q(ga_-).$$

It is obvious that $T_+(a)$ and $T_-(a)$ are bounded on A^+_n and $\overset{o}{A}{}^n_-$, respectively.

Definition. A decomposing algebra A is said to have the Fredholm property, if for every $a(t) \in GA_{n\times n}$ the following two conditions are satisfied:

(a) $T_+(a)$ and $T_-(a)$ are Fredholm operators on A^+_n and $\overset{o}{A}{}^n_-$, respectively.

(b) $\operatorname{Ind} T_+(a) = - \operatorname{Ind} T_-(a)$.

The Theorems 5.21 and 5.22 below will give simple sufficient conditions for a decomposing algebra to possess the Fredholm property.
The following theorem represents the main result of this chapter.

5.5. Theorem. Let A be a decomposing algebra. Then A has the factorization property if and only if A has the Fredholm property.

PRELIMINARIES

5.6. Let P_0 be the bounded operator defined on A or on $A_{n\times n}$ by $P_0 f = \chi_1 Q \chi_{-1} P f$. Furthermore, put

$$(f,g) = \frac{1}{2\pi i} \int_\Gamma g(t)f(t)dt;$$

this makes sense if either $f \in A_n$ and $g \in A^n$, or $f \in A_{n\times n}$ and $g \in A_{n\times n}$. Obviously, $P_0 f = (f, \chi_{-1} I)$, where I here and in what follows denotes the unit matrix of order n. On the other hand,

$$(f,g) = P_0(\chi_1 gf). \tag{1}$$

Now fix $g_- \in \overset{\circ}{A}{}^n_-$ and consider the linear functional $F(g_-)$ defined as follows:

$$F(g_-) : A_n^+ \to \mathbb{C}, \quad f_+ \to (f_+, g_-).$$

$F(g_-)$ is bounded due to (1). Moreover, $F(g_-) \neq 0$ for $g_- \neq 0$. In fact, assume $F(g_-)$ is the zero functional. Then $0 = (\chi_k I, g_-) = P_0 \chi_{k+1} g_-$ for $k = 0,1,2,\ldots$, which implies that $g_- = 0$.
Thus

$$F : \overset{\circ}{A}{}^n_- \to (A_n^+)^*, \quad g \mapsto F(g_-)$$

defines an embedding of $\overset{\circ}{A}{}^n_-$ into the adjoint space $(A_n^+)^*$ of A_n^+. We shall identify $\overset{\circ}{A}{}^n_-$ with the corresponding subspace of $(A_n^+)^*$. Analogously, A_n^+ may be regarded as subspace of $(\overset{\circ}{A}{}^n_-)^*$.

<u>5.7. Lemma.</u> The space $\overset{\circ}{A}{}^n_-$ is invariant under $T_+(a)^*$, and A_n^+ is invariant under $T_-(a)^*$. Moreover,

$$T_+(a)^* | \overset{\circ}{A}{}^n_- = T_-(a), \quad T_-(a)^* | A_n^+ = T_+(a). \tag{2}$$

<u>Proof.</u> If $a \in A_{n\times n}$, $f_+ \in A_n^+$, $g_- \in \overset{\circ}{A}{}^n_-$, then

$$g_-(Paf_+) = (Qg_-a)f_+ + Pg_-af_- - Qg_-Qaf$$

and from $P_0\chi_1 P_0 = 0$ and $P_0 \chi_1 Qg_-Q = 0$ we get in accordance with 5.6(1)

$$(T_+(a)f_+, g_-) = (f_+, T_-(a)g_-),$$

which implies (2) almost at once. Δ

PROOF OF NECESSITY

What we must prove is:

<u>5.8. Proposition.</u> If a decomposing algebra A has the factorization property, then it has also the Fredholm property.

<u>Proof.</u> Let $a(t) \in GA_{n\times n}$ and let $a = a_- d a_+$ be the factorization 5.2. From the definition of $T_\pm(a)$ we obtain

$$T_+(a) = T_+(a_-)T_+(d)T_+(a_+), \quad T_-(a) = T_-(a_+)T_-(d)T_-(a_-).$$

Clearly, $T_\pm(a_\pm)^{-1} = T_\pm(a_\pm^{-1})$, which yields the invertibility of $T_\pm(a_\pm)$. Furthermore, $T_+(d)$ and $T_-(d)$ are Fredholm, and thereby

$$\operatorname{Ind} T_+(d) = \sum_{j=1}^{n} \varkappa_j, \quad \operatorname{Ind} T_-(d) = -\sum_{j=1}^{n} \varkappa_j.$$

Hence $T_\pm(a)$ are also Fredholm and $\operatorname{Ind} T_+(a) = -\operatorname{Ind} T_-(a)$. Δ

We note a consequence of the Fredholm property which will be important in the following.

<u>5.9. Proposition.</u> Suppose A has the Fredholm property. Then for $a(t) \in GA_{n\times n}$

$$\operatorname{Ker} T_+(a)^* = \operatorname{Ker} T_-(a), \quad \operatorname{Ker} T_-(a)^* = \operatorname{Ker} T_+(a). \qquad (3)$$

In particular, $T_+(a)$ is invertible from the right (from the left) if and only if $T_-(a)$ is invertible from the left (from the right).

<u>Proof.</u> Identifying as in 5.6 $\overset{o}{A}{}^n_-$ with a part of $(A_n^+)^*$ and A_n^+ with a part of $(\overset{o}{A}{}^n_-)^*$ we get from 5.7(2)

$$\dim \operatorname{Ker} T_-(a) \leq \dim \operatorname{Ker} T_+(a)^* = \operatorname{codim} \operatorname{Im} T_+(a),$$

$$\dim \operatorname{Ker} T_+(a) \leq \dim \operatorname{Ker} T_-(a)^* = \operatorname{codim} \operatorname{Im} T_-(a).$$

Since $\operatorname{Ind} T_+(a) = -\operatorname{Ind} T_-(a) =: \varkappa$, we obtain

$$\dim \operatorname{Ker} T_+(a)^* = \dim \operatorname{Ker} T_+(a) - \varkappa$$

$$\leq \operatorname{codim} \operatorname{Im} T_-(a) - \varkappa = \dim \operatorname{Ker} T_-(a).$$

Thus $\dim \operatorname{Ker} T_+(a)^* = \dim \operatorname{Ker} T_-(a)$ and analogously, $\dim \operatorname{Ker} T_-(a)^* = \dim \operatorname{Ker} T_+(a)$. Combining this with 5.7(2) we arrive at the asserted equalities (3). Δ

CANONICAL FACTORIZATION

<u>5.10. Definition.</u> If $a(t) \in A_{n\times n}$ admits a factorization 5.2 with the identity as middle factor (i.e. $d(t) = I$), then the <u>factorization</u> is said to be <u>canonical.</u>

Clearly, if A is a decomposing algebra with factorization property, then $a(t) \in A_{n\times n}$ admits a canonical factorization if and only if $T_+(a)$ (or, by Proposition 5.9, $T_-(a)$) is invertible.

<u>5.11. Lemma.</u> Suppose A is a decomposing algebra with Fredholm property. Let $a(t) \in A_{n\times n}$ and let $T_+(a)$ be invertible. Then

there are $x_+(t) \in A^+_{n\times n}$ and $x_-(t) \in A^-_{n\times n}$ such that

$$a(t)x_+(t) = I + \overset{o}{y}_-(t), \quad x_-(t)a(t) = I + \overset{o}{y}_+(t), \tag{4}$$

where $\overset{o}{y}_-(t) \in \overset{o}{A}{}^-_{n\times n}$ and $\overset{o}{y}_+(t) \in \overset{o}{A}{}^+_{n\times n}$. Thereby the (constant) matrices $P_o x_+$ and $P_o x_-$ are regular.

<u>Proof.</u> Because of the invertibility of $T_+(a)$ there are $x_k^+(t) \in A_n^+$ such that $T_+(a)x_k^+ = e_k$, where e_k is the k th unit vector in $\mathbb{C}_n$. This is equivalent to $a(t)x_k^-(t) = e_k + \overset{o}{y}{}_k^-(t)$ for some $\overset{o}{y}{}_k^-(t) \in \overset{o}{A}{}_n^-$. Forming the matrices $x_+(t)$ and $\overset{o}{y}_-(t)$ with columns $x_k^+(t)$ and $\overset{o}{y}{}_k^-(t)$ we get the first equality in (4).

From Proposition 5.9 we obtain the invertibility of $T_-(a)$ and then in the same way as above the second equality in (4) can be derived. Let us now prove the regularity of $P_o x_+$. Write $x_+(t) = c + \overset{o}{x}_+(t)$, $c \in \mathbb{C}_{n\times n}$, $\overset{o}{x}_+(t) \in \overset{o}{A}{}^+_{n\times n}$. So $c = P_o x_+$. Let r be in $\mathbb{C}_n$ such that $cr = 0$. Then $t^{-1}x_+(t)r = t^{-1}\overset{o}{x}_+(t)r \in A^+_{n\times n}$ and from (4) we get

$$\begin{aligned} a(t)t^{-1}x_+(t)r &= t^{-1}a(t)x_+(t)r \\ &= t^{-1}I + t^{-1}\overset{o}{y}_-(t) \in \overset{o}{A}{}^-_{n\times n}. \end{aligned}$$

Hence $t^{-1}x_+(t)r \in \operatorname{Ker} T_+(a)$ and it results that $t^{-1}x_+(t)r = 0$, that is, $x_+(t)r = 0$. Again applying (4) yields $r + \overset{o}{y}_-(t)r = 0$, whence $r = P_o(r + \overset{o}{y}_-(t)r) = 0$. Thus $c = P_o x_+$ is regular.

The regularity of $P_o x_-$ can be proved equally. Δ

<u>5.12. Theorem.</u> Suppose A is a decomposing algebra with Fredholm property and let $a(t) \in A_{n\times n}$. If $T_+(a)$ is invertible, then $a(t)$ admits a canonical factorization.

<u>Remark.</u> This theorem contains the sufficiency portion of the main Theorem 5.5 for the case of canonical factorization. Splitting this part of the proof from the general case has only methodical reasons.

<u>Proof.</u> The equalities 5.11(4) give

$$r(t) := x_-(t)a(t)x_+(t) = x_-(t)(I+\overset{o}{y}_-(t)) = (I+\overset{o}{y}_+(t))x_+(t).$$

Since $A_+ \cap A_- = A_o$, we obtain $r(t) = P_o x_- = P_o x_+$. The regularity of $P_o x_\pm$ implies that of $r(t)$, therefore $x_\pm(t) \in GA^\pm_{n\times n}$ and the representation

$$a(t) = x_-^{-1}(t)(P_o x_+)x_+^{-1}(t)$$

is just a canonical factorization of $a(t)$. Δ

PROOF OF SUFFICIENCY

Now we are going to prove the following.

5.13. Proposition. If a decomposing algebra A has the Fredholm property, then it has also the factorization property.

Thus, suppose A is decomposing with Fredholm property and let $a(t) \in GA_{n\times n}$. Then $T_+(a)$ is Fredholm. Additionally, assume that

$$\operatorname{Ker} T_+(a) = \{0\}. \tag{5}$$

We shall see later that this is no restriction of generality.
To prove the above proposition we need some preliminaries, which are mainly taken from Heinig [2]. Set

$$C_k = \operatorname{Ker} T_-(a\chi_{-k}) \quad (k = -1,0,1,2,\ldots).$$

Obviously,

$$C_{-1} \supset C_0 = \operatorname{Ker} T_-(a) \supset C_1 \supset C_2 \supset \ldots$$

(note that we write $\supset$ to mean $\supseteqq$). If $u \in C_k$, $u \neq 0$, then $\chi_{-s}u \in \operatorname{Ker} T_-(a)$ for $s = 0,\ldots,k$. Thus $\dim \operatorname{Ker} T_-(a) \geqq k + 1$. Therefore $C_k = \{0\}$ for $k \geqq \dim \operatorname{Ker} T_-(a)$. Let q denote the largest index k with $C_k \neq \{0\}$.
A simple computation gives

$$u \in C_k \Longleftrightarrow T_-(a\chi_{-k-1})u \in \chi_{-1}A_0^n, \tag{6}$$

where, of course, $A_0^n = (A_0)^n$.

5.14. Lemma. Let c_k denote the codimension of C_{k+1} in C_k, that is, $c_k = \dim C_k/C_{k+1}$. Then

$$\dim C_k = \sum_{s=k}^{q} c_s, \tag{7}$$

$$n = c_{-1} \geqq c_0 \geqq c_1 \geqq \cdots \geqq c_q \geqq c_{q+1} = 0. \tag{8}$$

Proof. Equality (7) is obvious. Let us prove (8). Because of 5.13(5) and Proposition 5.9 the operator $T_-(a)$ is onto. Hence there are linear independent $x_i(t) \in \overset{0}{A}{}_-^n$ such that $T_-(a)x_i = \chi_{-1}e_i$ $(i=1,\ldots,n)$. From 5.13(6) we get $x_i \in C_{-1}$. A little thought shows

$$C_{-1} = C_0 \oplus \operatorname{lin}\{x_1,\ldots,x_n\},$$

where lin E denotes the linear span of a set E. Thus $c_{-1} = n$. By definition, $u \in C_{k+1}$ if and only if $u \in C_k$ and $\chi_{-1}u \in C_k$. Therefore,

$$C_{k+1} = \chi_1 C_k \cap C_k = \chi_1(C_k \cap \chi_{-1}C_k). \tag{9}$$

Since $\dim(A + B) = \dim A + \dim B - \dim(A \cap B)$, we obtain from (7)

$$\dim(C_k + \chi_{-1}C_k) = 2 \sum_{s=k}^{q} c_s - \sum_{s=k+1}^{q} c_s = \sum_{s=k}^{q} c_s + c_k. \qquad (10)$$

Now the inclusion

$$C_k + \chi_{-1}C_k \subset C_{k-1} \qquad (11)$$

gives $\sum_{s=k}^{q} c_s + c_k \leqq \sum_{s=k-1}^{q} c_s$, whence $c_k \leqq c_{k-1}$, which completes the proof. Δ

Put $\gamma_k = c_{k-1} - c_k$ $(k = 0,1,2,\dots,q+1)$. Then

$$\sum_{k=o}^{q+1} \gamma_k = n. \qquad (12)$$

Because of (10) and (11) we can find linear independent vector functions

$$u_{k,i}(t) \in \overset{o}{A}{}_-^n \quad (k = 0,\dots,q+1; \quad i = 1,\dots,\gamma_k) \qquad (13)$$

such that for $k = 0,\dots,q+1$

$$C_{k-1} = (C_k + \chi_{-1}C_k) \oplus U_k, \qquad (14)$$

where $U_k = \mathrm{lin}\{u_{k,1},\dots,u_{k,\gamma_k}\}$. In case $\gamma_k = 0$ we put $U_k = \{0\}$. By construction, for $u \in U_k$ $(k = 0,\dots,q+1)$,

$$T_-(a\chi_{-s})u = 0 \quad (s = -1,\dots,k-1), \qquad (15)$$

$$0 \neq T_-(a\chi_{-k})u \in \chi_{-1}A_o^n. \qquad (16)$$

<u>5.15. Lemma.</u> The vector functions

$$t^{-r}u_{l,i}(t) \quad (l = k+1,\dots,q+1; \ i=1,\dots,\gamma_l; \ r = 0,\dots,l-k-1) \qquad (17)$$

form a basis of C_k $(k = -1,\dots,q)$.

<u>Proof.</u> Let V_k be the set of vectors (17). Then it is easily seen that $\mathrm{lin}\, V_k \subset C_k$ and from 5.14(14) it follows that the elements of V_k are linear independent. Finally, from

$$\dim V_k = \gamma_{k+1} + 2\gamma_{k+2} + \dots + (q+1-k)\gamma_{q+1}$$
$$= c_k + c_{k+1} + \dots + c_q = \dim C_k$$

we obtain $\mathrm{lin}\, V_k = C_k$. Δ

Now form the matrix function $u_-(t) \in \overset{o}{A}{}^-_{n\times n}$ with the rows being $t^{-k}u_{k,i}(t)$ $(k = 0,\dots,q+1; \ i = 1,\dots,\gamma_k)$ (note that $\gamma_o+\dots+\gamma_{q+1}= n$

by 5.14(12)). From 5.14(15) we know that

$$u_-(t)a(t) = ht^{-1} + z_+(t), \tag{18}$$

where h is some constant matrix and $z_+(t) \in A^+_{n\times n}$. Note that h is regular. Indeed, if

$$r = (r_{01},\dots,r_{0\gamma_0}, r_{11},\dots,r_{1\gamma_1},\dots,r_{q+1,1},\dots,r_{q+1,\gamma_{q+1}})$$

is a constant row vector such that $rh = 0$, then

$$\sum_{k=0}^{q+1} \sum_{i=1}^{\gamma_q} r_{ki} t^{-k} u_{k,i}(t) = ru_-(t) \in \operatorname{Ker} T_-(a)$$

and the above lemma (with $k = -1$) implies $r = 0$.

5.16. Partial indices. For our further purposes it will be useful to replace the system of non-negative integers γ_k by another system $\{\varkappa_j\}$, which will be just the system of the partial indices appearing already in 5.2 (for the case where 5.13(5) holds!).

Definition. Let $\{\varkappa_j\}_{j=1}^n$, $\varkappa_1 \leqq \varkappa_2 \leqq \dots \leqq \varkappa_n$, be the system of integers which contains exactly γ_0 times the number 0, exactly γ_1 times the number 1,..., exactly γ_{q+1} times the number $q+1$ (note that $\gamma_0 + \dots + \gamma_{q+1} = n$ by 5.14(12)). Thus, by convention, k does not belong to the system $\{\varkappa_j\}$ if $\gamma_k = 0$.
Formally,

$$\{\varkappa_j\} = \{\underbrace{0,\dots,0}_{\gamma_0}, \underbrace{1,\dots,1}_{\gamma_1},\dots, \underbrace{q+1,\dots,q+1}_{\gamma_{q+1}}\} .$$

The system $\{\varkappa_j\}$ is called the system of partial indices.

The function $u_{k,i}(t)$ introduced by 5.14(13) will now be denoted by $u_j(t)$, where the index j is defined to be $j = \gamma_0 + \dots + \gamma_{k-1} + i$. Thus, the system

$$\{u_{01},\dots,u_{0\gamma_1}, u_{11},\dots,u_{1\gamma_1},\dots,u_{q+1,1},\dots,u_{q+1,\gamma_{q+1}}\}$$

takes the form

$$\{u_1,\dots,u_{\gamma_1}, u_{\gamma_1+1},\dots,u_{\gamma_1+\gamma_2},\dots,u_n\}.$$

Now form the matrix function $x_-(t) \in A^-_{n\times n}$ whose j th row is just $tu_j(t)$. Then 5.15(18) can be written in the form

$$d_-(t)x_-(t)a(t) = h + \overset{\circ}{y}_+(t),$$

where h is a constant regular matrix, $\overset{\circ}{y}_+ = \chi_1 z_+ \in \overset{\circ}{A}{}^+_{n\times n}$ and $d_-(t) = \operatorname{diag}(t^{-\varkappa_1},\dots,t^{-\varkappa_n})$.

5.17. Lemma. Let $x_-(t) \in A^-_{n\times n}$ be the matrix function introduced in 5.16. Then $P_0 x_-$ is a regular (constant) matrix.

Proof. Assume $r = (r_1,\dots,r_n) \neq 0$ is a row vector such that $rP_0x_- = 0$ and let l denote the smallest index with $r_l \neq 0$. By 5.16 the j th row of $x_-(t)t^{-1}$ is just $u_j(t)$. Define $w(t) = rx_-(t)t^{-1}$. Since $rP_0x_- = 0$, we get $w(t) \in \overset{o}{A}{}^n_-$ and, obviously,

$$w(t) = \sum_{j=l}^{n} r_j u_j(t). \tag{19}$$

Hence

$$\begin{aligned} w(t)a(t)t^{-\varkappa_l} &= t^{-\varkappa_l} \sum_{j=l}^{n} r_j u_j(t)a(t) \\ &= \sum_{j=l}^{n} t^{\varkappa_j-\varkappa_l} r_j t^{-\varkappa_j} u_j(t)a(t) \\ &= rf_+(t)d_-(t)t^{-1}x_-(t)a(t) \\ &= t^{-1}rf_+(t)d_-(t)x_-(t)a(t) \\ &= t^{-1}rf_+(t)(h + \overset{o}{\vartheta}_+(t)) \in A^n_+ , \end{aligned}$$

where, by definition,

$$f_+(t) = \operatorname{diag}(\underbrace{0,\dots,0,}_{l-1}\ t^{\varkappa_l-\varkappa_l},\dots,t^{\varkappa_n-\varkappa_l}) \in A^+_{n\times n}.$$

This implies $w \in C_{\varkappa_l}$ and by Lemma 5.15 we have

$$w(t) = \sum_{s,k,m} \alpha_{s,k,m} t^{-s} u_{k,m}(t), \tag{20}$$

where the sum runs through

$$k = \varkappa_l+1,\dots,q+1; \quad m = 1,\dots,\gamma_k; \quad s = 0,\dots,k-\varkappa_l-1.$$

On comparing (20) with (19) we see that $\alpha_{s,k,m} = 0$ for $s \neq 0$, that is,

$$w(t) = \sum_{k,m} \beta_{k,m} u_{k,m}(t), \tag{21}$$

and the sum runs through $k = \varkappa_l+1,\dots,q+1$; $m = 1,\dots,\gamma_k$. Because of 5.16 the sum (21) may be written as

$$w(t) = \sum \beta_i u_i(t) \quad (i = \gamma_0+\dots+\gamma_{\varkappa_l}+1,\dots,n). \tag{22}$$

But

$$\gamma_0+\dots+\gamma_{\varkappa_l}+1 \geqq \gamma_{\varkappa_1} + \gamma_{\varkappa_2}+\dots+\gamma_{\varkappa_l} + 1 \geqq l+1, \tag{23}$$

since $\gamma_{\varkappa_j} \geqq 1$. By combining (19), (22) and (23) we find that $r_l = 0$, which contradicts to our assumption $r_1 \neq 0$. Thus $P_o x_-$ is regular. Δ

5.18. Lemma. There exist matrix functions $x_\pm \in A^{\pm}_{n\times n}$ such that

$$d_-(t)x_-(t)a(t) = h + \overset{o}{y}_+(t), \tag{24}$$

$$a(t)x_+(t)d_-(t) = g + \overset{o}{y}_-(t), \tag{25}$$

where $d_-(t) = \mathrm{diag}(t^{-\varkappa_1},\ldots,t^{-\varkappa_n})$ with the $\varkappa_j$'s defined in 5.16, $\overset{o}{y}_\pm \in \overset{o}{A}{}^{\pm}_{n\times n}$ and $h,g,P_o x_-,P_o x_+$ are regular (constant) matrices.

Proof. For $x_-(t)$ the assertion follows from 5.16 and 5.17. Let us prove the existence of an $x_+(t)$ having the desired properties. The row vector functions $u_j(t) \in \overset{o}{A}{}^n_-$ defined in 5.14, 5.17 can be regarded as linear functionals on A^+_n (cf. 5.6). By construction they are linear independent. We claim their linear independence on the subspace $(A_o)_n \subset A^+_n$. In fact, assume $u(t) = \sum_{j=1}^{n} \alpha_j u_j(t) \in \overset{o}{A}{}^n_-$ vanishes on $(A_o)_n$. Then by 5.6(1) $P_o tu(t) = 0$, i.e. $tu(t) \in \overset{o}{A}{}^n_-$. Now, again by 5.6(1), $P_o t(tu(t)) = 0$, that is, $t^2u(t) \in \overset{o}{A}{}^n_-$ etc. Thus $t^k u(t) \in \overset{o}{A}{}^n_-$ for all $k = 0,1,2,\ldots$, which implies $u(t) \equiv 0$ in contradiction to the linear independence of the u_j's on A^n_+. So the functions u_j are linear independent on $(A_o)_n$ and, consequently, there exists a basis $\{g_j\}$ $(j = 1,\ldots,n)$ of $(A_o)_n$ such that $(g_j,u_k) = \delta_{jk}$ (δ the Kronecker symbol).
Next we want to show that g_j is orthogonal to $C_{\varkappa_j}$. From Lemma 5.15 we know a basis $\{t^{-r}u_{k,i}(t)\}$ of $C_{\varkappa_j}$. Obviously,

$$(g_j,t^{-r}u_{k,i}(t)) = P_o t^{-r+1}u_{k,i}(t)g_j = 0$$

for $r \geqq 1$. On the other hand, by 5.16 the $u_{k,i}$'s with $k = \varkappa_j+1,\ldots,q+1;$ $i = 1,\ldots,\gamma_k$ are just the u_l's with $l \geqq \gamma_o + \ldots+\gamma_{\varkappa_j} + 1 \geqq j+1$ (see 5.17(23)). This gives $(g_j,u_l) = 0$, which proves the orthogonality of g_j to $C_{\varkappa_j}$.
Now, $C_{\varkappa_j}$ is just the kernel of $T_-(a\chi_{-\varkappa_j})$ and by Proposition 5.9 we have $C_{\varkappa_j} = \mathrm{Ker}\, T_+(a\chi_{-\varkappa_j})^*$. Since $T_+(a\chi_{-\varkappa_j})$ is Fredholm, we get

$$C_{\varkappa_j} \oplus \mathrm{Im}\, T_+(a\chi_{-\varkappa_j}) = A^+_n.$$

Hence $g_j \in \mathrm{Im}\, T_+(a\chi_{-\varkappa_j})$, which implies the existence of $v_j(t) \in A^+_n$ such that $T(a\chi_{-\varkappa_j})v_j = g_j$, or

$$a(t)v_j(t)t^{-\varkappa_j} = g_j + \overset{o}{y}{}^-_j(t),$$

where $\overset{o}{y}_j^-(t) \in \overset{o}{A}_n^-$. To obtain (25) it only remains to form the matrices $x_+(t)$, g and $\overset{o}{y}_-(t)$ having as their columns $v_j(t)$, g_j and $\overset{o}{y}_j^-(t)$, respectively. Δ

5.19. Proof of Proposition 5.13. Suppose A is a decomposing algebra with Fredholm property and let $a(t) \in GA_{n\times n}$. Assume for a moment that 5.13(5) holds. We apply Lemma 5.18 and set

$$p(t) = d_-(t)x_-(t)a(t)x_+(t).$$

Then

$$p(t) = (h+\overset{o}{y}_+(t))x_+(t) = d_-(t)x_-(t)(g+\overset{o}{y}_-(t))d_-^{-1}(t). \tag{26}$$

Therefore $p(t) \in A^+_{n\times n}$ and $p(t)d_-(t) \in A^-_{n\times n}$. Hence $p(t)$ is a matrix polynomial (in t). We write

$$x_-(t)(g+\overset{o}{y}_-(t)) = (c_{jk}(t))_{j,k=1}^n, \; p(t) = (p_{jk}(t))_{j,k=1}^n.$$

Then $p_{jk}(t) = t^{\varkappa_k - \varkappa_j} c_{jk}(t)$. Since $p_{jk}(t) \in A_+$ and $c_{jk}(t) \in A_-$, we see that $p_{jk}(t) = 0$ for $\varkappa_k < \varkappa_j$ and $p_{jk}(t) = \text{const.}$ for $\varkappa_k = \varkappa_j$. Therefore $p(t)$ has a triangular block-structure

$$\begin{pmatrix} p_1 & * & \cdots & * \\ 0 & p_2 & \cdots & * \\ \vdots & \vdots & & \vdots \\ 0 & 0 & \cdots & p_l \end{pmatrix}$$

where the p_i's are constant matrices. A little thought shows that the p_i's are submatrices of $(P_0x_-)\cdot g$. By Lemma 5.18, p_i is regular, which implies that $p(t)$ has constant non-vanishing determinant. Thus, $p(t)^{-1}$ is also a matrix polynomial (in t). From (26) we get $x_+(t)^{-1} = p(t)^{-1}(h + \overset{o}{y}_+(t)) \in A^+_{n\times n}$. Analogously can be proved that $x_-(t)^{-1} \in A^-_{n\times n}$. Thus $a(t) = a_-(t)d(t)a_+(t)$ with

$$a_-(t) = x_-(t)^{-1}, \quad d(t) = d_-(t)^{-1}, \quad a_+(t) = p(t)x_+(t)^{-1}$$

is a factorization in the sense of 5.2.

Finally, let us remove the extra assumption 5.13(5). To that end put $\alpha = \dim \operatorname{Ker} T_+(a)$. As in 5.13 one can show that $T_+(a\chi_k)$ has only a trivial kernel for $k \geqq \alpha$. Thus all the above arguments apply to $b(t) = a(t)t^\alpha$ and, consequently, $b(t)$ admits a factorization $b(t) = a_-(t)d_-(t)a_+(t)$. Then

$$a(t) = a_-(t) \cdot t^{-\alpha}d_-(t) \cdot a_+(t)$$

represents the desired factorization of $a(t)$. Δ

Now we are going to state some sufficient conditions for a decomposing algebra A to have the Fredholm property.
We begin by noting that the operator $T_+(a)$ may be identified with the Toeplitz operator $T(a)$ acting on A_n^+. The following proposition will establish the connection between $T_-(a)$ and a certain Toeplitz operator.

5.20. Proposition. Suppose A is a decomposing algebra and let $a \in GA_{n\times n}$. Then $T_-(a)$ is Fredholm on $\overset{\circ}{A}{}_-^n$ if and only if $T[(a^\tau)^{-1}]$ is Fredholm on A_n^+. In that case

$$\operatorname{Ind} T_-(a) = \operatorname{Ind} T[(a^\tau)^{-1}].$$

Here $a^\tau \in A_{n\times n}$ denotes the transposed matrix function of $a \in A_{n\times n}$.

Proof. A little thought shows that via the natural isomorphism of $\overset{\circ}{A}{}_-^n$ and $\overset{\circ}{A}{}_n^-$ the operator $T_-(a)$ may be identified with the operator $QM(a^\tau)Q|\overset{\circ}{A}{}_n^-$. But

$$(P + QM(a^\tau)Q)(I + PM(a^\tau)Q) = P + M(a^\tau)Q =$$

$$= M(a^\tau)(PM[(a^\tau)^{-1}]P + Q)(I + QM[(a^\tau)^{-1}]P)$$

and the operators $M(a^\tau)$, $I + PM(a^\tau)Q$, $I + QM[(a^\tau)^{-1}]P$ are invertible. Thus, $P + QM(a^\tau)Q$ is Fredholm if and only if $Q + PM[(a^\tau)^{-1}]P$ is, and in this case the kernel and cokernel dimensions of these operators coincide. This proves the proposition. Δ

5.21. Theorem. Denote by R the set of all rational functions without poles on Γ and let A be a decomposing algebra. If one of the sets

$$(R \cap \overset{\circ}{A}_-) + A_+ \quad \text{or} \quad \overset{\circ}{A}_- + (R \cap A_+)$$

is dense in A, then A has the Fredholm property.

Proof. Suppose $(R \cap \overset{\circ}{A}_-) + A_+$ is dense in A. We assert that $H(\tilde{a}) \in \mathfrak{C}_\infty(A_+)$ for every $a \in A$. Indeed, given $a \in A$, we can choose a sequence $\{a_k\} \subset (R \cap \overset{\circ}{A}_-) + A_+$ such that $\|a - a_k\|_A \to 0$ as $k \to \infty$. By Kronecker's Lemma (cf. 2.17(a)) $H(\tilde{a}_k)$ is finite-rank and therefore $H(\tilde{a})$ being uniform limit of finite-rank operators is compact. Thus

$$H(\tilde{a}) \in \mathfrak{C}_\infty(A_+) \quad \text{for every} \quad a \in A. \tag{27}$$

Let now $a = (a_{ij})_{i,j=1}^n \in GA_{n\times n}$. By (27) the (operator) entries of the matrix $(T(a_{ij}))_{i,j=1}^{n}$ commute modulo $\mathfrak{C}_\infty(A_+)$. Hence the corollary in 1.15 and Proposition 2.7 may be applied to conclude that

$$T(a) \in \Phi(A_n^+) \iff T(\det a) \in \Phi(A_+).$$

But the Fredholmness of $T(\det a)$ follows from the invertibility of $\det a$ and (27): a regularizer is $T[(\det a)^{-1}]$.
Our next concern is the equality

$$\operatorname{Ind} T(a) = \operatorname{Ind} T(\det a). \tag{28}$$

We approximate again $a \in A_{n\times n}$ by $a_k \in [(R \cap \overset{o}{A}_-) + A_+]_{n\times n}$. Then $a_k \in GA_{n\times n}$ for all sufficiently large k and the above argument gives $T(a_k) \in \Phi(A_n^+)$ and $T(\det a_k) \in \Phi(A_+)$. But $H(\tilde{a}_k)$ is of finite-rank and so by 1.16

$$\operatorname{Ind} T(a_k) = \operatorname{Ind} T(\det a_k). \tag{29}$$

On passing to the limit $k \to \infty$ in (29) the equality (28) results. Similarly can be shown that

$$T[(a^\tau)^{-1}] \in \Phi(A_n^+), \quad T[\det((a^\tau)^{-1})] \in \Phi(A_+),$$

$$\operatorname{Ind} T[(a^\tau)^{-1}] = \operatorname{Ind} T[\det((a^\tau)^{-1})].$$

To complete the proof we are left with the verification of

$$\operatorname{Ind} T[\det a] = - \operatorname{Ind} T[\det((a^\tau)^{-1})].$$

But this is obvious, since $\det((a^\tau)^{-1}) = (\det a)^{-1}$ and therefore $T[\det((a^\tau)^{-1})]$ is a regularizer of $T[\det a]$. Δ

5.22. Theorem. Let A and B be algebras with the following properties:

(i) B is a decomposing algebra with Fredholm property.

(ii) A is densely and continuously embedded in B.

(iii) For every $a \in A$ the Hankel operator $H(a)$ (or $H(\tilde{a})$) maps B_+ into A_+.

(iv) For every $a \in GA_{n\times n}$ the Toeplitz operator $T(a)$ is Fredholm on A_n^+.

Then A is a decomposing algebra with Fredholm property.

Proof. To conclude that A is decomposing combine the relation $Pa - \chi_1 H(a)\chi_0 \in A_0$ (or $(Qa)^\sim = H(\tilde{a})\chi_0$) with (i), (ii), (iii).

It remains to show the Fredholm property. Let $a \in GA_{n\times n}$. We assert that the kernels of $T(a)$ on A_n^+ and B_n^+ coincide.

Indeed, let $T(a)\varphi = 0$, $\varphi \in B_n^+$. Then

$$T(a^{-1})T(a)\varphi = \varphi - H(a^{-1})H(\tilde{a})\varphi = 0,$$

consequently, $\varphi = H(a^{-1})H(\tilde{a})\varphi$ and from (iii) it results that $\varphi \in A_n^+$. Similarly one can prove that the kernels of $T(a)^*$ on $(A_n^+)^*$ and $(B_n^+)^*$. coincide.

Next if $a \in GA_{n\times n}$ then $a \in GB_{n\times n}$ by (ii). Owing to (i) and Proposition 5.20

$$T(a) \in \Phi(B_n^+), \quad T[(a^\tau)^{-1}] \in \Phi(B_n^+),$$

$$\operatorname{Ind} T(a) = - \operatorname{Ind} T[(a^\tau)^{-1}] \quad (\text{in } B_n^+). \tag{30}$$

Due to (iv) and Proposition 5.20

$$T(a) \in \Phi(A_n^+), \quad T[(a^\tau)^{-1}] \in \Phi(A_n^+)$$

and since kernel and cokernel dimensions of these operators on A_n^+ and B_n^+ coincide, we can conclude from (30) that

$$\operatorname{Ind} T(a) = - \operatorname{Ind} T[(a^\tau)^{-1}] \quad (\text{in } A_n^+),$$

which completes the proof. Δ

5.23. Examples of algebras with factorization (Fredholm) property.

(a) The Wiener algebra W (see 2.2). This is a consequence of Theorem 5.21 and the fact that $\operatorname{lin}\{\chi_k : k \in \mathbb{Z}\}$ is dense in W.

(b) The weighted Wiener algebras $W^{\alpha,\beta}$ $(\alpha \geqq 0, \beta \geqq 0)$.

Here $W^{\alpha,\beta}$ consists, by definition, of all functions on Γ whose Fourier coefficients satisfy

$$\|a\|_{W^{\alpha,\beta}} = \sum_{j=-\infty}^{-1} |a_j|(|j| + 1)^\alpha + \sum_{j=0}^{\infty} |a_j|(j+1)^\beta < \infty.$$

That $W^{\alpha,\beta}$ has the Fredholm property follows as in (a).

(c) The algebras $W \cap l_2^{\alpha,\beta}$ $(\alpha \geqq 0, \beta \geqq 0)$ (see 4.13).

First of all we show that $W \cap l_2^{\alpha,\beta}$ is actually an algebra. To that end let a,b be in $W \cap l_2^{\alpha,\beta}$, put $a_+ = Pa$, $a_- = Qa$, $b_+ = Pb$, $b_- = Qb$ and write $ab = a_+b_+ + a_+b_- + a_-b_+ + a_-b_-$. Then $a_+b_+ \in l_2^\beta$ (cf. 4.13(12)). This holds for $0 < \beta < 1$ due to Lemma 4.17, and for $\beta \geqq 1$ (even for $\beta > 1/2$) owing to 4.13(b). Analogously can be shown that $\tilde{a}_-\tilde{b}_- \in l_2^\alpha$.

To prove that a_+b_- and a_-b_+ are in $l_2^{\alpha,\beta}$ we remark only that for $c_- \in W \cap \overline{H^\infty}$ the Toeplitz operator $T(c_-)$ is bounded on l_2^γ $(\gamma \geqq 0)$; this is easily seen from

$$\|T(c_-)\|_{\mathcal{B}(l^\gamma_2)} \leq \sum_{n=0}^{\infty} |c_{-n}| \|T(\chi_{-n})\|_{\mathcal{B}(l^\gamma_2)}$$

and $\|T(\chi_{-n})\|_{\mathcal{B}(l^\gamma_2)} = 1.$

The Fredholm property of $W \cap l_2^{\alpha,\beta}$ is now an obvious consequence of Theorem 5.21.

(d) The algebras $K_{p,q}^{\alpha,\beta}$ in the case where $p > 1$, $q > 1$, $1/p + 1/q = 1$, $\alpha > 0$, $\beta > 0$, $\alpha + \beta = 1$, and $\alpha \neq 1/p$ (equivalently, $\beta \neq 1/q$) (see 4.15).

Above all, $K_{p,q}^{\alpha,\beta}$ is decomposing by virtue of Corollary 2 in 4.16. Next, there is no loss of generality in assuming that $\beta > 1/q$. Since the trigonometric polynomials are dense in each Besov space B_q^β, $q < \infty$ (see Bergh, Löfström [1], 6.2.4) and since B_q^β is continuously embedded in C for $\beta > 1/q$ (cf. Lemma 4.16), we can conclude that $R \cap (B_q^\beta)_A$ is dense in B_q^β. It remains to apply Theorem 5.21.

(e) The algebras $W \cap K_{p,q}^{1/p,1/q}$ $(1 \leq p < \infty,\ 1 \leq q < \infty)$ (see 4.10).
By (a), any matrix function $a \in (W \cap K_{p,q}^{1/p,1/q})_{n\times n}$ admits a factorization $a = a_- d_- d_+ a_+$, where $a_+ \in G(W \cap H^\infty)_{n\times n}$, $a_- \in G(W \cap \overline{H^\infty})_{n\times n}$, $d_-(t) = \operatorname{diag}(t^{-\lambda_1},\dots,t^{-\lambda_n})$, $d_+(t) = \operatorname{diag}(t^{\mu_1},\dots,t^{\mu_n})$, λ_i and μ_i $(i = 1,\dots,n)$ are non-negative integers. It remains to show that a_- and a_+ are in $(K_{p,q}^{1/p,1/q})_{n\times n}$ (cf. Theorem 4.11). But

$$d_-^{-1}a_-^{-1} \cdot (Pa) + d_-^{-1}a_-^{-1} \cdot (Qa) = d_+a_+,$$

whence $P[d_-^{-1}a_-^{-1} \cdot (Pa)] = d_+a_+$, so that $d_+a_+ \in (B_q^{1/q})_A$ due to Lemma 4.18. Similarly can be shown that $\tilde{d}_-\tilde{a}_- \in (B_p^{1/p})_A$.

(f) The algebras c^γ $(0 < \gamma < 1)$ (see 1.20).
This is immediate from Theorem 5.21.

(g) The algebras C^γ $(0 < \gamma < 1)$ (see 1.20).
We put $A = C^\gamma$ and $B = c^{\gamma-\varepsilon}$, where ε is an arbitrary real number between 0 and γ, and apply Theorem 5.22. Condition (i) is fulfilled by virtue of (f). For (ii) we refer to Prößdorf, Silbermann [1], p.40. Finally, (iii) and (iv) follow from the well-known fact that $H(a) \in \mathfrak{C}_\infty(C^\delta, C^\gamma)$ for $a \in C^\gamma$ and arbitrary δ, $0 < \delta < 1$ (see e.g. Prößdorf [1], p.103). Another possibility to verify (iii) is the following: if $a \in C^\gamma$ and $\varphi_+ \in H^\infty$ then $T(\tilde{\varphi}_+) \in \mathcal{B}(C^\gamma)$ by Lemma 4.18 $(p = \infty)$ and therefore $H(a)\varphi_+ = T(\tilde{\varphi}_+)(P(\chi_{-1}a)) \in C^\gamma$.

(h) The algebras $c^{\gamma} + H^{\infty}$ and $C^{\gamma} + H^{\infty}$ $(0 < \gamma < 1)$ (see 4.21). This is straightforward for $c^{\gamma} + H^{\infty}$ from Theorem 5.21 and follows for $C^{\gamma} + H^{\infty}$ in the same way as in (g).

5.24. Toeplitz operators with symbols in W. In chapter 2 the case $p = 1$ has played an exceptional role. Actually, however, this case turns out to be much simpler than the case $p > 1$. This is due to the fact that $M_1 = W$ (see 2.2) and therefore almost all properties of Toeplitz operators on l^1 can be derived from the Wiener-Hopf factorization.

Of course, also in the case $p > 1$ the tool of Wiener-Hopf factorization can successfully be applied to the study of Toeplitz operators once we know that the symbol belongs to W. The following theorem summatrizes some facts about Toeplitz operators with symbols in W.

Theorem. Let $1 \leq p < \infty$ and suppose $a \in W_{N \times N}$. Then $T(a)$ is a Fredholm operator (Φ_+- or Φ_--operator) on l^p_N if and only if $\det a(t) \neq 0$ for all $t \in \Gamma$. In this case $\operatorname{Ind} T(a) = - \operatorname{ind} \det a$, $a \in GW_{N \times N}$ and $T(a^{-1})$ is a regularizer of $T(a)$. If

$$a(t) = a_-(t) \cdot \operatorname{diag}(t^{\varkappa_1}, \ldots, t^{\varkappa_N}) \cdot a_+(t)$$

is a factorization in the sense of 5.2, then

$$\dim \operatorname{Ker} T(a) = - \sum_{\varkappa_j < 0} \varkappa_j,$$

$$\operatorname{codim} \operatorname{im} T(a) = \sum_{\varkappa_j > 0} \varkappa_j.$$

In particular, $T(a) \in G\mathcal{B}(l^p_N)$ if and only if $\varkappa_1 = \ldots = \varkappa_N = 0$. In this case $T(a_+^{-1})\, T(a_-^{-1})$ is the inverse of $T(a) = T(a_-)T(a_+)$.

Proof. The sufficiency part is in view of 5.23(a) almost obvious and the necessity portion can be proved by a perturbation argument (see Remark 2 in 2.21). Δ

NOTES AND COMMENTS

For the long history of the Wiener-Hopf factorization, its different methods, various results, and many applications we refer the reader to the books Muskhelishvili [1], Gohberg, Krupnik [2], Gohberg, Feldman [1], Prößdorf [1], Michlin, Prößdorf [1], and especially we recommend the recent monograph Clancey, Gohberg [2].

The approach presented in the text is due to Heinig, Silbermann [1]. We should emphasize that the factorization founded here has simultaneously the following two properties:

(a) It does not go out of the algebra under consideration.

(b) This algebra may contain certain discontinuous functions.

Throughout this chapter the integer n (the matrix-dimension) has been supposed to be fixed. Thus, it would be more precise to speak about the n-factorization property and the n-Fredholm property. Then Theorem 5.5 asserts that a decomposing algebra has the n-factorization property if and only if it has the n-Fredholm property. Clearly, it would be of great interest to know whether every decomposing algebra with 1-factorization property has also the n-factorization property. Although all the concrete examples considered in 5.23 indicate that this seems to be so, we have not been able to give an answer to this question for the general case.

C H A P T E R 6

TOEPLITZ DETERMINANTS

For a function $a \in L^{\infty}_{N\times N}$ consider the determinant $D_n(a)$ of the finite Toeplitz matrix $T_n(a)$ (this is an $(n+1)N \times (n+1)N$-matrix) generated by $a(t)$,

$$D_n(a) = \det T_n(a) = \det(a_{j-k})^n_{j,k=0} \quad (n = 0,1,2,\dots).$$

If $a \in H^{\infty}_{N\times N}$ or $a \in \overline{H^{\infty}_{N\times N}}$ then $T_n(a)$ has triangular block structure and therefore $D_n(a) = (\det a_0)^{n+1}$. It is a rather surprising fact that even under much more general conditions (say $a \in C^{1/2+\varepsilon}_{N\times N}$, $\varepsilon > 0$, $T(a)$ and $T(\tilde{a})$ invertible on l^2_N) the determinants have still an asymptotic behavior of the form

$$D_n(a) \sim G(a)^{n+1}E(a) \quad (n \to \infty),$$

or equivalently,

$$\log D_n(a) = (n+1)\log G(a) + \log E(a) + o(1) \quad (n \to \infty),$$

where $G(a)$ and $E(a)$ are certain constants. Thus one could ask for an expansion of the form

$$\log D_n(a) = (n+1)\log G(a) + \log E(a) +$$

$$+ \sum_{k=1}^{M} \frac{\alpha_k}{(n+1)^k} + O\left(\frac{1}{(n+1)^{M+1}}\right) \quad (n \to \infty).$$

The somewhat unexpected result is: if $a(t)$ is very smooth, say $a \in C^{(M+2)/2}$, and if $T(a)$ and $T(\tilde{a})$ are invertible on l^2_N, then $\alpha_1 = \dots = \alpha_M = 0$. This problem and some other questions will be discussed in the present chapter.

THE TERMS G(a) AND E(a)

<u>6.1. Proposition.</u> If $a \in L^{\infty}_{N\times N}$ and $D_n(a) \neq 0$ then

$$D_{n-1}(a)/D_n(a) = \det P_0 T_n^{-1}(a) P_0.$$

<u>Proof.</u> For $N = 1$ this follows from Cramer's rule and for general N from Jacobi's theorem on minors of the inverse matrix (see Gantmacher [1], p. 20). Δ

<u>Corollary.</u> If $a \in L^{\infty}_{N\times N}$ and $T(a) \in \Pi\{P_n\}$ (that is, $T(a) \in \Pi_2\{P_n\}$) then

$$\lim_{n\to\infty} D_{n-1}(a)/D_n(a) = \det P_0 T^{-1}(a) P_0.$$

There is an interesting expression for $\det P_0 T^{-1}(a) P_0$ in case $a \in W_{N\times N}$. In fact, if $T(a)$ is invertible on l^2_N then by virtue of 5.24 we have a canonical factorization $a = a_- a_+$ in the sense of 5.2, 5.10 and therefore $T^{-1}(a) = T(a_+^{-1})T(a_-^{-1})$. Thus

$$\det P_0 T^{-1}(a) P_0 = \det P_0 T(a_+^{-1}) P_0 T(a_-^{-1}) P_0$$

$$= \det T_0(a_+^{-1}) \cdot \det T_0(a_-^{-1}).$$

By identifying functions in $W \cap H^\infty$ for a moment with their analytic extension into $\mathbb{D}$ we have

$$\det T_0(a_+^{-1}) = \det(a_+^{-1})_0 = \det(a_+^{-1}(0)) = 1/\det(a_+(0)) =$$

$$= 1/(\det a_+)(0) = 1/\exp \log(\det a_+)(0) =$$

$$= 1/\exp(\log \det a_+)(0) = 1/\exp(\log \det a_+)_0$$

and similarly

$$\det \; T_0(a_-^{-1}) = 1/\exp(\log \det a_-)_0.$$

Therefore,

$$\det P_0 T^{-1}(a) P_0 = 1/\exp\{(\log \det a_+)_0 + (\log \det a_-)_0\} =$$

$$= 1/\exp\{\log(\det a_- \cdot \det a_+)\}_0$$

$$= 1/\exp(\log \det a)_0.$$

To find an analogous expression in case $a \in (C + H^\infty)_{N\times N}$ will be our next concern.

6.2. Let $b \in G(C + H^\infty)$ and $\operatorname{ind} b = 0$ (see 1.19). We denote by $\hat{b}$ the harmonic extension of b into $\mathbb{D}$ and set $b_r(e^{i\vartheta}) = \hat{b}(re^{i\vartheta})$ for $0 \leqq r < 1$, $0 \leqq \vartheta < 2\pi$. Thus $b_r \in W$. By our hypotheses, there exists a δ, $0 < \delta < 1$, such that for $1-\delta < r < 1$ we have $b_r(t) \neq 0$ for all $t \in \Gamma$ and $\operatorname{ind} b_r = 0$. Consequently, there is a continuously defined $\log b_r$, and it can easliy by shown that $\log b_r$ as $r \to 1-0$ converges in the L^1-norm to a determination of $\log b$.

6.3. Definition. Let $a \in G(C + H^\infty)_{N\times N}$ and suppose $\operatorname{ind} \det a = 0$. Then $G(a)$ is defined as

$$G(a) = \lim_{r\to 1-0} \exp(\log(\det a)_r)_0.$$

In particular, if $a \in GC_{N\times N}$ and $\operatorname{ind} \det a = 0$ then

$$G(a) = \exp(\log \det a)_0.$$

6.4. Proposition. If $a \in G(C + H^{\infty})_{N\times N}$ and if $T(a)$ and $T(\tilde{a})$ are invertible on l^2_N then

$$\det P_0 T^{-1}(a) P_0 = 1/G(a). \tag{1}$$

Remark. We shall see in Theorem 6.6 that this proposition remains true even if $T(\tilde{a})$ is not invertible. Recall in this connection Theorem 2.33 and Proposition 2.34.

Proof. For $a \in W_{N\times N}$ this has already been proved in 6.1. Now take a general $a \in G(C + H^{\infty})_{N\times N}$ and define $a_r \in W_{N\times N}$ via the harmonic extension in analogy to 6.2. Since $G(a) = \lim_{r\to 1-0} G(a_r)$ by definition, to obtain (1) it suffices to show that $T(a_r) \in G\mathcal{B}(l^2_N)$ for r sufficiently close to 1 and that $T^{-1}(a_r)$ converges strongly to $T^{-1}(a)$. To that end let us prove that

$$H[(\tilde{a}_r)^{-1}] - H[\widetilde{a}^{-1}] = o(1) \quad (r \to 1-0), \tag{2}$$

where $o(1)$ denotes any sequence of operators tending to zero in the operator norm. By the asymptotic multiplicativity of the mapping $\tilde{a} \mapsto \tilde{a}_r$ (see 1.19) we have

$$\|(\widetilde{a}^{-1})_r - (\tilde{a}_r)^{-1}\|_{L^{\infty}} \to 0 \quad (r \to 1-0)$$

and thus, (2) is equivalent to

$$H(b_r) - H(b) = o(1) \qquad (r \to 1-0),$$

where $b = \widetilde{a}^{-1} \in (C + H^{\infty})_{N\times N}$. But this is trivial for $b \in \chi_j \overline{H^{\infty}}$ $(j = 0,1,2,\ldots)$, and follows for general b from the density of the union of these spaces in $(C + H^{\infty})_{N\times N}$. Thus (2) is proved and therefore

$$H[a_r]H[(\tilde{a}_r)^{-1}] - H[a]H[\widetilde{a}^{-1}]$$

$$= H[a_r](H[(\tilde{a}_r)^{-1}] - H[\widetilde{a}^{-1}]) + (H[a_r] - H[a])H[\widetilde{a}^{-1}]$$

$$= o(1) \quad (r \to 1-0),$$

since $H[a_r]$ converges strongly to $H[a]$ and the operator $H[\widetilde{a}^{-1}]$ is compact by Theorem 2.16 so that 1.3(d) applies. From Proposition 2.7 we get

$$T(a_r)T(a_r^{-1}) = T(a)T(a^{-1}) + o(1) \quad (r \to 1-0). \tag{3}$$

Since $T(a)T(a^{-1})$ is invertible and since $T(a_r)$ and $T(a_r^{-1})$ are both Fredholm of index zero, we obtain that actually $T(a_r)$ and $T(a_r^{-1})$ themselves are invertible. Moreover, from (3) it follows

$$T^{-1}(a_r^{-1})T^{-1}(a_r) = T^{-1}(a^{-1})T^{-1}(a) + o(1) \quad (r \to 1-0),$$

hence

$$T^{-1}(a_r) = T(a_r^{-1})T^{-1}(a^{-1})T^{-1}(a) + o(1) \quad (r \to 1-0),$$

which converges strongly to $T^{-1}(a)$. Δ

To remove the extra assumption of the invertibility of $T(\tilde{a})$ we need the following theorem.

6.5. Theorem (Widom). Let $a \in L^\infty_{N\times N}$ and suppose $T(a)$ is Fredholm on l^2_N and has index zero. Then there is a matrix function φ with only finitely many non-vanishing Fourier coefficients such that $T(a + \varepsilon\varphi)$ is invertible on l^2_N for all ε belonging to some sufficiently small punctured disk with center $\varepsilon = 0$.

For lack of place we refer the reader for a proof to Widom [7].

6.6. Theorem. If $a \in G(C + H^\infty)_{N\times N}$ and $T(a) \in G\mathcal{B}(l^2_N)$ then

$$\det P_0T^{-1}(a)P_0 = 1/G(a). \tag{4}$$

Proof. From Theorem 2.35 it follows that $T(\tilde{a})$ is Fredholm of index zero. By the previous theorem there is a trigonometric matrix polynomial φ such that $T(\tilde{a} + \varepsilon\tilde{\varphi}) \in G\mathcal{B}(l^2_N)$ for all $\varepsilon \in C$ satisfying $0 < |\varepsilon| < r$. Replacing if necessary r by a smaller number, we have also $T(a + \varepsilon\varphi) \in G\mathcal{B}(l^2_N)$ for $0 \leq |\varepsilon| < r$. Thus

$$\det P_0T^{-1}(a + \varepsilon\varphi)P_0 = 1/G(a + \varepsilon\varphi) \tag{5}$$

for $0 < |\varepsilon| < r$. Now $1/G(a + \varepsilon\varphi)$ is analytic in the full disk $|\varepsilon| < r$ as uniform limit of analytic functions and $\det P_0T^{-1}(a+\varepsilon\varphi)P_0$ is analytic in some disk with center $\varepsilon = 0$ since $T^{-1}(a + \varepsilon\varphi)$ can be expanded into a power series. Thus both sides of (5) are analytic in some disk with center $\varepsilon = 0$, therefore equality holds for $\varepsilon = 0$, too. Δ

6.7. Proposition. If $a \in G(C + H^\infty)_{N\times N}$ and $T(a) \in G\mathcal{B}(l^2_N)$ then for $n = 0,1,2,\ldots$

$$\det P_nT^{-1}(a)P_n = 1/G(a)^{n+1}.$$

Proof. For $a \in W_{N\times N}$ the usual factorization argument applies, that is,

$$\det P_nT^{-1}(a)P_n = \det P_nT(a_+^{-1})P_nT(a_-^{-1})P_n =$$
$$= \det T_n(a_+^{-1})\cdot\det T_n(a_-^{-1}) = (\det T_0(a_+^{-1})\cdot\det T_0(a_-^{-1}))^{n+1} =$$
$$= (\det P_0T^{-1}(a)P_0)^{n+1} = 1/G(a)^{n+1}$$

(note that $T_n(a_{\pm}^{-1})$ have triangular block structure) and for general $a(t)$ we may proceed as in 6.4 and 6.6. Δ

Thus we have evaluated the determinants of the finite sections of $T^{-1}(a)$. But under certain conditions $T(a)$ differs from $T^{-1}(a^{-1})$ only by a trace class operator. So the role played by the following lemma becomes clear.

<u>6.8. Lemma.</u> If $A \in \mathcal{GB}(l_N^2)$, $A \in \Pi\{P_n\}$, and $K \in \mathfrak{C}_1(l_N^2)$ then

$$\lim_{n\to\infty} \frac{\det P_n(A + K)P_n}{\det P_nAP_n} = \det(I + A^{-1}K).$$

<u>Proof.</u> Put $A_n = P_nAP_n|\operatorname{Im} P_n$ and $K_n = P_nKP_n$. By Proposition 3.3 the matrices A_n are invertible for all sufficiently large n and $A_n^{-1}P_n$ converges strongly to A^{-1}. From 1.3(d) we obtain that $\|A_n^{-1}P_nKP_n - A^{-1}K\|_1 \to 0$ as $n \to \infty$ and keeping in mind 1.8(d),(a) we get

$$\frac{\det(A_n + K_n)}{\det A_n} = \det(P_n + A_n^{-1}K_n) = \det(I + A_n^{-1}K_n) =$$

$$= \det(I + A_n^{-1}P_nKP_n) \to \det(I + A^{-1}K) \ (n \to \infty). \ \Delta$$

<u>Corollary.</u> Let $a \in G(C + H^{\infty})_{N\times N}$ and suppose $T(a) \in \mathcal{GB}(l_N^2)$, $T(\tilde{a}) \in \mathcal{GB}(l_N^2)$, $H(a)H(\widetilde{a^{-1}}) \in \mathfrak{C}_1(l_N^2)$. Then $T(a)T(a^{-1}) - I \in \mathfrak{C}_1(l_N^2)$ and

$$\lim_{n\to\infty} D_n(a)/G(a)^{n+1} = \det T(a)T(a^{-1}).$$

<u>Proof.</u> We apply the preceding lemma with $A = T^{-1}(a^{-1})$ and $A + K = T(a)$. Note that

$$K = - H(a)H(\widetilde{a^{-1}})T^{-1}(a^{-1}) \in \mathfrak{C}_1(l_N^2) \qquad (6)$$

by our hypotheses combined with 2.34(18), therefore $A \in \Pi\{P_n\}$ by Theorem 3.7.
From Proposition 6.4 we get

$$\det P_nT^{-1}(a^{-1})P_n = 1/G(a^{-1})^{n+1}$$

and since $\|(\det a^{-1})_r - (\det a)_r^{-1}\|_{L^\infty} \to 0$ as $r \to 1 - 0$ by virtue of the asymptotic multiplicativity (see 1.19), we have $G(a^{-1}) = 1/G(a)$, which gives

$$\det P_nT^{-1}(a^{-1})P_n = G(a)^{n+1}.$$

Finally,

$$I + A^{-1}K = I + T(a^{-1})[T(a)-T^{-1}(a^{-1})] = T(a^{-1})T(a),$$

which in view of $\det T(a^{-1})T(a) = \det T(a)T(a^{-1})$ (see 1.8(c)) completes the proof. Δ

<u>6.9. Definition.</u> Let $a \in GL^{\infty}_{N\times N}$ and suppose $T(a)T(a^{-1}) - I = H(a)H(\widetilde{a^{-1}})$ is a trace class operator. Then $E(a)$ is defined as

$$E(a) = \det T(a)T(a^{-1}).$$

From 1.8(f) it follows that $E(a) \neq 0$ if and only if both $T(a)$ and $T(a^{-1})$, or equivalently, both $T(a)$ and $T(\tilde{a})$ are invertible on l^2_N.

<u>6.10. Theorem.</u> Suppose

$$p > 1, \quad q > 1, \; 1/p + 1/q = 1, \quad \alpha > 0, \; \beta > 0, \; \alpha + \beta = 1$$

and

$$0 < |\alpha - 1/p| < 1/2 \quad (\text{equivalently, } 0 < |\beta - 1/q| < 1/2).$$

Let $a \in GK^{\alpha,\beta}_{p,q}$ and assume $\operatorname{ind} a = 0$ (cf. 4.24). Then

$$T(a)T(a^{-1}) - I \in \mathfrak{C}_1(l^2) \tag{7}$$

and with $\log a$ defined as in 6.2,

$$E(a) = \exp \sum_{k=1}^{\infty} k(\log a)_k(\log a)_{-k}. \tag{8}$$

We emphasize that this is a theorem for the scalar case.

<u>Proof.</u> That $T(a)T(a^{-1}) - I$ belongs to $\mathfrak{C}_1(l^2)$ has already been stated in 4.15. Due to 5.23(d) we have a factorization $a = a_-a_+$, where $a_+ \in G(B^{\beta}_q \cap H^{\infty})$ and $a_- \in G(B^{\alpha}_p \cap \overline{H^{\infty}})$.

As a consequence of Theorem 4.24, the spectrum of a_+ as an element in $B^{\beta}_q \cap H^{\infty}$ coincides with its spectrum as element in H^{∞}. But the latter spectrum is known from 1.18 to be clos $\hat{a}_+(\mathbb{D})$ and applying a standard argument (cf. for instance Krein [2]) we can conclude that there is a determination of $\log a_+$ in $B^{\beta}_q \cap H^{\infty}$.
Similarly can be shown that there exists a determination of $\log a_-$ in $B^{\alpha}_p \cap \overline{H^{\infty}}$. Consequently, $\log a_- + \log a_+$ is a determination of $\log a$ in $K^{\alpha,\beta}_{p,q}$ coinciding up to an additive constant with that given in 6.2.
Now a few application of Proposition 2.7 gives

$$\begin{aligned} T(a)T(a^{-1}) &= T(a_-)T(a_+)T(a_-^{-1})T(a_+^{-1}) \\ &= e^{T(\log a_-)}e^{T(\log a_+)}e^{-T(\log a_-)}e^{-T(\log a_+)} \,. \end{aligned}$$

By construction, $T(\log a_{\pm}) \in \mathfrak{B}(l^2)$ and

$$T(\log a_-)T(\log a_+) - T(\log a_+)T(\log a_-)$$

$$= H(\log a_+)H(\log \tilde{a}_-) \in \mathfrak{C}_1(l^2)$$

so that by virtue of 1.10

$$\det T(a)T(a^{-1}) = \exp \operatorname{tr} H(\log a_+)H(\log \tilde{a}_-)$$

$$= \exp \sum_{k=1}^{\infty} k(\log a_+)_k(\log a_-)_{-k}$$

$$= \exp \sum_{k=1}^{\infty} k(\log a)_k(\log a)_{-k}. \quad \Delta$$

Remark 1. The conclusion (7) and (8) of this theorem can by shown to remain true if $a(t)$ is required to be an invertible element of index zero in the algebra $W \cap K_{p,q}^{1/p,1/q}$, where $p > 1$, $q > 1$, $1/p + 1/q = 1$. The arguments to prove this are in view of 5.23(e) essentially the same as in the preceding proof.

Remark 2. It is a rather subtle fact that the above theorem is valid for $p = q = 2$, $\alpha = \beta = 1/2$. This was proved in Widom [8]. The theorem seems to hold for $\alpha = 1/p$, $\beta = 1/q$ $(p \neq 2)$, too, but we are not able to prove this.

Remark 3. It appears to be an open question what the constant $E(a) = \det T(a)T(a^{-1})$ looks like in the matrix case (even for very smooth symbols).

THE STRONG SZEGÖ LIMIT THEOREM

6.11. Theorem. Let $a \in (K_{p,q}^{\alpha,\beta})_{N\times N}$ where $p > 1$, $q > 1$, $1/p + 1/q = 1$, $\alpha > 0$, $\beta > 0$, $\alpha + \beta = 1$, $-1/2 < \alpha - 1/p < 1/2$ $(-1/2 < \beta - 1/q < 1/2)$. If $T(a)$ and $T(\tilde{a})$ are invertible on l_N^2 then

$$\lim_{n\to\infty} D_n(a)/G(a)^{n+1} = E(a). \tag{9}$$

Proof. From the invertibility of $T(a)$ and $T(\tilde{a})$ we deduce the invertibility of $a(t)$ in $(K_{p,q}^{\alpha,\beta})_{N\times N}$ by Theorems 2.35 and 4.24. To complete the proof it suffices to recall 4.15 and the corollary in 6.8. Δ

Remark 1. The theorem remains true under the conditions that $a \in (K_{0,1}^{0,1})_{N\times N}$ or $a \in (K_{1,0}^{1,0})_{N\times N}$ and that $T(a) \in G\mathfrak{B}(l_N^2)$ and $T(\tilde{a}) \in G\mathfrak{B}(l_N^2)$. In this case $H(a) \in \mathfrak{C}_1(l_N^2)$ or $H(\widetilde{a^{-1}}) \in \mathfrak{C}_1(l_N^2)$ and so the above argument applies.

Remark 2. Theorem 6.14 below will show that it is sufficient to require only $T(a) \in \Phi(l^2_N)$ and $\operatorname{Ind} T(a) = 0$.

We give still another proof of Theorem 6.11, which will be based on the identity 3.7(13). For this purpose we need a lemma.

6.12. Lemma. Let $a \in L^\infty_{N\times N}$ and $T(a) \in G\mathcal{B}(l^2_N)$. Then $P_nT^{-1}(a)P_n : \operatorname{Im} P_n \to \operatorname{Im} P_n$ is invertible for all $n = 0,1,2,\ldots$ and

$$(P_nT^{-1}(a)P_n)^{-1}P_n = W_nT^{-1}(\widetilde{a}^{-1})W_n.$$

Proof. First of all, $T(\widetilde{a}^{-1}) \in G\mathcal{B}(l^2_N)$ by 2.33, 2.34 (or as an immediate consequence of Proposition 3.8). Application of 3.8 and 3.6(6), (7) gives

$$\begin{aligned}
&(P_nT^{-1}(a)P_n)^{-1}P_n \\
&= P_nT(a)P_n - P_nT(a)Q_n(Q_nT(a)Q_n)^{-1}Q_nT(a)P_n \\
&= P_nT(a)P_n - W_nH(\widetilde{a})T(\chi_{-n})(Q_nT(a)Q_n)^{-1}T(\chi_n)H(a)W_n \\
&= W_n(T(\widetilde{a})-H(\widetilde{a})T^{-1}(a)H(a))W_n = W_nT^{-1}(\widetilde{a}^{-1})W_n. \ \Delta
\end{aligned}$$

6.13. Here another proof of Theorem 6.11. We shall write $o_1(1)$ to denote any sequence of operators converging in the $\mathcal{C}_1$-norm to zero. Since

$$T^{-1}(a)-T(a^{-1}) = T(a^{-1})H(a)H(\widetilde{a}^{-1}) \in \mathcal{C}_1(l^2_N),$$

$$T^{-1}(\widetilde{a})-T(\widetilde{a}^{-1}) = T(\widetilde{a}^{-1})H(\widetilde{a})H(a^{-1}) \in \mathcal{C}_1(l^2_N),$$

and Q_n converges strongly to zero as $n \to \infty$, we obtain from 3.7(13) $R_nT_n(a) = P_n + o_1(1)$, where

$$R_n = P_nT^{-1}(a)P_n + W_n(T^{-1}(\widetilde{a})-T(\widetilde{a}^{-1}))W_n.$$

But $T(a) \in \Pi\{P_n\}$ by our hypotheses, therefore $T_n^{-1}(a) = R_n + o_1(1)$. Replacing $T^{-1}(a)-T(a^{-1})$ by K and using Lemma 6.12 we get

$$\begin{aligned}
T_n^{-1}(a) &= R_n + o_1(1) \\
&= (P_nT^{-1}(a)P_n)\{P_n+(P_nT^{-1}(a)P_n)^{-1}W_nKW_n + o_1(1)\} \\
&= (P_nT^{-1}(a)P_n)\{P_n+W_nT^{-1}(\widetilde{a}^{-1})P_nKW_n + o_1(1)\},
\end{aligned}$$

thus by Proposition 6.7

$$\frac{1}{D_n(a)} = \frac{1}{G(a)^{n+1}} \det\{I + P_nT^{-1}(\widetilde{a}^{-1})P_nKP_n + o_1(1)\}$$

and because of 1.3(d) and 1.8(a), (c) we have

$$\det\{I + P_nT^{-1}(\widetilde{a}^{-1})P_nKP_n + o_1(1)\}$$

$$\to \det\{I + T^{-1}(\widetilde{a}^{-1})(T^{-1}(\widetilde{a})-T(\widetilde{a}^{-1}))\}$$

$$= \det T^{-1}(\widetilde{a}^{-1})T^{-1}(\widetilde{a}) = 1/\det T(\widetilde{a})T(\widetilde{a}^{-1})$$

$$= 1/\det T(a)T(a^{-1}),$$

which completes the proof. Δ

<u>6.14. Theorem.</u> Let $a \in (K^{\alpha,\beta}_{p,q})_{N\times N}$ where $p > 1$, $q > 1$, $1/p + 1/q = 1$, $\alpha > 0$, $\beta > 0$, $\alpha + \beta = 1$, $-1/2 < \alpha - 1/p < 1/2$ $(-1/2 < \beta - 1/q < 1/2)$. If $T(a)$ is Fredholm on l^2_N and has index zero then

$$\lim_{n\to\infty} D_n(a)/G(a)^{n+1} = E(a). \tag{10}$$

<u>Proof.</u> Define

$$\Psi = \{a \in (K^{\alpha,\beta}_{p,q})_{N\times N} : T(a),\ T(\widetilde{a}) \in G\mathcal{B}(l^2_N)\},$$

$$\Phi = \{a \in (K^{\alpha,\beta}_{p,q})_{N\times N} : T(a) \in \Phi(l^2_N),\ \mathrm{Ind}\ T(a) = 0\}.$$

Note that $a \in \Phi$ implies $T(\widetilde{a}) \in \Phi(l^2_N)$ and $\mathrm{Ind}\ T(\widetilde{a}) = 0$ by Theorem 2.35. Thus we can deduce from Theorem 6.5 that Ψ forms an open and dense subset of Φ.
Now take $a \in \Phi$. Then there is an $\varepsilon > 0$ such that $b \in \Phi$ for all $b \in (K^{\alpha,\beta}_{p,q})_{N\times N}$ satisfying $\|b - a\|_{(K^{\alpha,\beta}_{p,q})_{N\times N}} < \varepsilon$. Among these b's we can find a $b_o \in \Psi$. Put $a_\zeta = (1 - \zeta)a + \zeta b_o$. Then $a_\zeta \in \Phi$ for all ζ in an open set F containing the disk $|\zeta| \leqq 1$ and by 1.8(b) the function $g(\zeta) = \det T(a_\zeta)T(a_\zeta^{-1})$ is analytic in F.
Moreover, $g(1) \neq 0$ by 1.8(f). Therefore, the set of zeros of $g(\zeta)$ is discrete, and thus, there is an r with $0 < r < 1$ such that $g(\zeta) \neq 0$ for $|\zeta| = r$. Consequently, $a_\zeta \in \Psi$ for all ζ on the circle with radius r. By what has already been shown,

$$\lim_{n\to\infty} D_n(a_\zeta)\ G(a_\zeta)^{n+1} = g(\zeta) \tag{11}$$

for $|\zeta| = r$. Moreover, a check of the proof given in 6.13 shows that this holds uniformly on the circle. It follows that (11) holds throughout the disk $|\zeta| \leqq 1$, in particular at $\zeta = 0$ and this is what was wanted. Δ

6.15. We know from Proposition 6.1 that

$$D_{n-1}(a)/D_n(a) = \det P_n T_n^{-1}(a) P_0.$$

We begin with writing $P_0 T_n^{-1}(a) P_0$ in a form which will be favourable for our further investigation. To that end we make the following assumptions:

(a) There are two factorizations $a = u_- u_+ = v_+ v_-$ where $u_+, v_+ \in GH^\infty_{N\times N}$ and $u_-, v_- \in G\overline{H^\infty_{N\times N}}$.

(b) $u_- \in C_{N\times N}$ or $u_+ \in C_{N\times N}$.

Thus $a \in (C + H^\infty)_{N\times N}$ or $a \in (C + \overline{H^\infty})_{N\times N}$, the operators $T(a)$, $T(\tilde{a})$, $T(a^{-1})$ etc. are invertible on l^2_N and $T(a) \in \Pi\{P_n\}$ by Theorem 3.7. Now Proposition 3.8 gives

$$P_0 T_n^{-1}(a) P_0 = P_0 T^{-1}(a) P_0 - P_0 T^{-1}(a) Q_n (Q_n T^{-1}(a) Q_n)^{-1} Q_n T^{-1}(a) P_0$$

$$= P_0 T(u_+^{-1}) P_0 \{I - P_0 T(u_-^{-1}) Q_n (Q_n T^{-1}(a) Q_n)^{-1} Q_n T(u_+^{-1}) P_0\} P_0 T(u_-^{-1}) P_0.$$

Our next concern is the term $(Q_n T^{-1}(a) Q_n)^{-1} Q_n$. Because of

$$T^{-1}(a) = T(u_+^{-1}) T(u_-^{-1}) = T(u_+^{-1} u_-^{-1}) - H(u_+^{-1}) H(\tilde{u}_-^{-1}) =$$

$$= T(a^{-1}) - K, \quad K \in \mathfrak{C}_\infty(l^2_N),$$

we have

$$(Q_n T^{-1}(a) Q_n)^{-1} Q_n = (Q_n T(a^{-1}) Q_n - Q_n K Q_n)^{-1} Q_n =$$

$$= (A_n - B_n)^{-1} Q_n = (I - A_n^{-1} B_n)^{-1} A_n^{-1} Q_n,$$

where $A_n = Q_n T(a^{-1}) Q_n | \operatorname{Im} Q_n$ and $B_n = Q_n K Q_n$. Obviously,

$$A_n^{-1} Q_n = Q_n T(v_+) Q_n T(v_-) Q_n$$

and $\|B_n\|_\infty \to 0$ as $n \to \infty$ (see 1.3(d)). Expansion into Neumann's series yields

$$(Q_n T^{-1}(a) Q_n)^{-1} Q_n = \sum_{k=0}^{\infty} (A_n^{-1} B_n)^k A_n^{-1} Q_n,$$

therefore,

$$P_0 T_n^{-1}(a) P_0 = P_0 T(u_+^{-1}) P_0 \{I - \sum_{k=0}^{\infty} G_{n,k}\} P_0 T(u_-^{-1}) P_0 \tag{12}$$

where

$$G_{n,o} = P_o T(u_-^{-1}) Q_n A_n^{-1} Q_n T(u_+^{-1}) P_o = P_o T(c) Q_n T(b) P_o, \quad (13)$$

$$G_{n,k} = P_o T(u_-^{-1}) Q_n (A_n^{-1} B_n)^k Q_n A_n^{-1} Q_n T(u_+^{-1}) P_o$$

$$= P_o T(c)(Q_n H(b) H(\tilde{c}) Q_n)^k T(b) P_o \quad (k \geqq 1), \quad (14)$$

$$b = v_- u_+^{-1}, \quad c = u_-^{-1} v_+. \quad (15)$$

In a similar fashion it can be shown that

$$T_n^{-1}(a) = P_n T(u_+^{-1}) P_n \{I - \sum_{k=o}^{\infty} F_{n,k}\} P_n T(u_-^{-1}) P_n, \quad (16)$$

where

$$F_{n,o} = P_n T(c) Q_n T(b) P_n (= W_n H(\tilde{c}) H(b) W_n), \quad (17)$$

$$F_{n,k} = P_n T(c)(Q_n H(b) H(\tilde{c}) Q_n)^k T(b) P_n \quad (k \geqq 1), \quad (18)$$

with b and c again given by (15). On account of

$$\det P_n T(u_+^{-1}) P_n \cdot \det P_n T(u_-^{-1}) P_n$$

$$= \det P_n T(u_+^{-1}) T(u_-^{-1}) P_n = \det P_n T^{-1}(a) P_n = 1/G(a)^{n+1}$$

we get for all sufficiently large n from (12), (16)

$$\frac{D_{n-1}(a)}{D_n(a)} = \frac{1}{G(a)} \det\{I - \sum_{k=o}^{\infty} G_{n,k}\}, \quad (19)$$

$$\frac{1}{D_n(a)} = \frac{1}{G(a)^{n+1}} \det\{I - \sum_{k=o}^{\infty} F_{n,k}\}. \quad (20)$$

6.16. Since $\det \exp A = \exp \operatorname{tr} A$ for a finite square matrix A, we obtain

$$\det\{I - \sum_{k=o}^{\infty} G_{n,k}\} = \exp \operatorname{tr} \log\{I - \sum_{k=o}^{\infty} G_{n,k}\}.$$

Now suppose we have a decomposition

$$\operatorname{tr} \log\{I - \sum_{k=o}^{\infty} G_{n,k}\} = - \operatorname{tr} H_n + s_n,$$

where $\{H_n\}$ is some sequence of matrices and $\sum_{n=o}^{\infty} |s_n| < \infty$. From

$$D_n(a) = \frac{D_n(a)}{D_{n-1}(a)} \cdots \frac{D_{n_o+1}(a)}{D_{n_o}(a)} D_{n_o}(a) \quad (21)$$

we get

$$\frac{D_n(a)}{G(a)^{n+1}} \exp\{-\mathrm{tr}(H_1+\ldots+H_n)\}\exp\{-(s_{n+1}+s_{n+2}+\ldots)\}$$
$$= \tilde{E} = \mathrm{const.} \neq 0,$$

or equivalently,

$$\log D_n(a) = (n+1)\log G(a) + \mathrm{tr}(H_1+\ldots+H_n) +$$
$$+ \log \tilde{E}(a) + (s_{n+1}+s_{n+2}+\ldots).$$

6.17. Suppose the conditions 6.15(a),(b) are satisfied and in addition assume we know that $\lim\limits_{n\to\infty} D_n(a)/G(a)^{n+1} = E(a)$ (this holds for instance if the hypotheses of Theorem 6.11 are fulfilled). Then 6.15(19) combined with 6.16(21) leads to the following interesting identity (n sufficiently large):

$$D_n(a) = G(a)^{n+1}E(a) \prod_{j=n+1}^{\infty} \det\{I - \sum_{k=0}^{\infty} G_{j,k}\}.$$

6.18. Theorem. Suppose the matrix function $a(t)$ satisfies one of the following conditions:

(i) $a \in W^{\alpha,\beta}_{N\times N}$.

(ii) $a = u_-u_+$, $u_+ \in G(C^{\alpha} \cap H^{\infty})_{N\times N}$, $u_- \in G(C^{\beta} \cap \overline{H^{\infty}})_{N\times N}$.

Furthermore, assume $T(a)$ and $T(\tilde{a})$ are invertible on l^2_N. Define $H_k(a)$ for $k = 1,2,3,\ldots$ as

$H_k(a) = 0$ if $\alpha + \beta > 1$;

$H_k(a) = G_{k,0}$ if $\alpha + \beta > 1/2$;

$H_k(a) = G_{k,0}+G_{k,1} + G^2_{k,1}/2$ if $\alpha + \beta > 1/3$.

(see 6.15(13),(14)). Then

$$\lim_{n\to\infty} \frac{D_n(a)}{G(a)^{n+1}} \exp\{-\mathrm{tr}(H_1(a)+\ldots+H_n(a))\} = \tilde{E}(a), \qquad (22)$$

or equivalently,

$$\log D_n(a) = (n+1)\log G(a) + \log \tilde{E}(a)$$
$$+ \mathrm{tr}(H_1(a)+\ldots+H_n(a)) + o(1) \quad (n \to \infty) \qquad (23)$$

with some constant $\tilde{E}(a) \neq 0$. Moreover, $o(1)$ in (23) can be replaced by

$o(1/n^{(\alpha+\beta)p-1})$ if (i) is satisfied with $\alpha + \beta > 1/p$, $p \in \mathbb{Z}_+$;

$O(1/n^{(\alpha+\beta)p-1})$ if (ii) is satisfied with $\alpha + \beta > 1/p$, $p \in \mathbb{Z}_+$;

and we have $\operatorname{tr} H_n(a) = o(1/n^{\alpha+\beta})$ and $\operatorname{tr} H_n(a) = O(1/n^{\alpha+\beta})$ in the case (i) and (ii), respectively. In particular, for $\alpha + \beta > 1$ the formulas (22), (23) are true with

$$H_1(a) = \ldots = H_n(a) = 0, \quad \tilde{E}(a) = E(a),$$

where $E(a)$ is defined by 6.9.

Remark 1. The prove given below shows how the $H_k(a)$'s are to be defined for $\alpha + \beta > 1/4$, $\alpha + \beta > 1/5$ etc. in order that (22), (23) hold.

Remark 2. The case $\alpha = \beta$ is perhaps of greatest interest. Then (ii) may by replaced by the condition $a \in C^{\alpha}_{N\times N}$.

Remark 3. A simple computation shows that

$$G_{n,o} = \sum_{k=1}^{\infty} c_{-n-k} b_{n+k},$$

$$G_{n,1} = \sum_{i,j,k=1}^{\infty} c_{-n-i} b_{n+i+j} c_{-n-j-k} b_{n+k},$$

where c_l and b_l are the matrix Fourier coefficients of b and c (see 6.15(15)).

Remark 4. This theorem makes clear what is meant by "effect of smoothness". If $\alpha + \beta$ is large $(\alpha + \beta > 1)$ then (22), (23) give higher order correction terms and estimates for the remainder in 6.11(9). On the other hand, if $\alpha + \beta$ is small $(\alpha + \beta < 1)$ then (22) shows how by adding some auxiliary terms on the left hand side the existence of certain limits involving Toeplitz determinants can be ensured. This is what we call "non-classical" Szegö limit theorems.

Proof. In the case (i) we have a factorization $a = u_- u_+ = v_+ v_-$, where $u_+, v_+ \in G(W^{\alpha,\beta} \cap H^{\infty})_{N\times N}$ and $u_-, v_- \in G(W^{\alpha,\beta} \cap \overline{H^{\infty}})_{N\times N}$ (see 5.23(b)). Equally $a = v_+ v_-$, where $v_+ \in GH^{\infty}_{N\times N}$ and $v_- \in G\overline{H^{\infty}_{N\times N}}$ for the case (ii) (see 5.23(g)). Thus the arguments of 6.15, 6.16 apply.

Under the condition (i), the matrix functions $b = v_- u_+^{-1}$ and $c = u_-^{-1} v_+$ are in $W^{\alpha,\beta}_{N\times N}$. Therefore as $n \to \infty$,

$$\|Q_n T(b) P_o\| = \|Q_n T(Pb - P_n Pb) P_o\| \leqq \|Pb - P_n Pb\|_W =$$

$$= \sum_{k=n+1}^{\infty} |b_k| \leqq (n+1)^{-\beta} \sum_{k=n+1}^{\infty} |b_k| (k+1)^{\beta} = o(n^{-\beta}),$$

$$\|Q_n H(b)\| = \|Q_n H(Pb - P_n Pb)\| \leqq \|Pb - P_n Pb\|_W = o(n^{-\beta})$$

and similarly $\|P_oT(c)Q_n\| = o(n^{-\alpha})$, $\|H(\tilde{c})Q_n\| = o(n^{-\alpha})$. This gives

$$\|G_{n,k}\| = o(n^{-(\alpha+\beta)(k+1)}) \quad (n \to \infty).$$

As concerns the case (ii) so notice that

$$Q_nT(b)P_o = Q_nT(v_-)Q_nT(u_+^{-1})P_o,$$

$$Q_nH(b) \quad = Q_nT(v_-)Q_nH(u_+^{-1})$$

and if p_n denotes the polynomial of degree n (in t) of best uniform approximation to u_+^{-1}, then $\|u_+^{-1} - p_n\|_{L^\infty} = O(n^{-\beta})$, whence

$$\|Q_nT(u_+^{-1})P_o\| = \|Q_nT(u_+^{-1}-p_n)P_o\| \leqq \|u_+^{-1}-p_n\|_{L^\infty} = O(n^{-\beta}),$$

$$\|Q_nH(u_+^{-1})\| = \|Q_nH(u_+^{-1}-p_n)\| \leqq \|u_+^{-1}-p_n\|_{L^\infty} = O(n^{-\beta})$$

and analogously for $P_oT(c)Q_n$ and $H(\tilde{c})Q_n$. Thus

$$\|G_{n,k}\| = O(n^{-(\alpha+\beta)(k+1)}) \quad (n \to \infty).$$

To complete the proof write

$$\operatorname{tr} \log\{I - \sum_{k=o}^{\infty} G_{n,k}\} = -\operatorname{tr} \sum_{k=o}^{\infty} G_{n,k} - \operatorname{tr}(\sum_{k=o}^{\infty} G_{n,k})^2/2 - \dots ,$$

expand $(\sum_{k=o}^{\infty} G_{n,k})^l$ $(l = 2,3,\dots)$, and proceed as in 6.16. Δ

6.19. Lemma. Suppose b and c satisfy one of following conditions:

(i) $b,c \in W^{\alpha,\beta}$

(ii) $b = u_-u_+$, $c = v_-v_+$; $u_+,v_+ \in G(C^\beta \cap H^\infty)$, $u_-,v_- \in G(C^\alpha \cap \overline{H^\infty})$.

Then if $\alpha + \beta > 1/p$, $p \in \mathbb{Z}_+$,

$$H(\tilde{c})H(b) \in \mathfrak{C}_p(l^2).$$

Proof. Assume b and c are in $W^{\alpha,\beta}$. Then $\dim \operatorname{Im} H(P_nPb) \leqq n+1$, consequently, we have for the s-numbers of $H(b)$ (cf. 1.2) as $n \to \infty$

$$s_{n+1}(H(b)) \leqq \|H(b)-H(P_nPb)\| \leqq \|Pb-P_nPb\|_W = o(n^{-\beta})$$

(see the proof in 6.18) and analogously $s_{n+1}(H(\tilde{c})) = o(n^{-\alpha})$. Thus by Horn's lemma (quoted in 1.2)

$$\sum_{n=1}^{N} s_n^p(H(\tilde{c})H(b)) = o(\sum_{n=1}^{\infty} n^{-(\alpha+\beta)p}) = O(1) \quad (N \to \infty),$$

that is, $H(\tilde{c})H(b) \in \mathfrak{C}_p(l^2)$.

If (ii) is fulfilled then also as in 6.18,

$$s_{n+1}(H(b)) = O(n^{-\beta}),\ s_{n+1}(H(\tilde{c})) = O(n^{-\alpha}) \quad (n \to \infty),$$

and from 1.3(d) it follows

$$\|\sum_{k=1}^{\infty} F_{n,k}\|_1 = o(1) \quad (n \to \infty).$$

So because of 1.8(a)

$$\det_1(I - F_{n,o}) - \det_1(I - \sum_{k=o}^{\infty} F_{n,k}) = o(1) \quad (n \to \infty).$$

Thus

$$\begin{aligned}
\frac{G(a)^{n+1}}{D_n(a)} &= \det_1(I - \sum_{k=o}^{\infty} F_{n,k}) \\
&= \det_1(I - F_{n,o}) + o(1) \\
&= \det_1 W_n(I - H(\tilde{c})H(b))W_n + o(1) \\
&= \det_1 P_n(I - H(\tilde{c})H(b))P_n + o(1) \\
&= \det_1 (I - H(\tilde{c})H(b)) + o(1) \\
&= \det_1 T(\tilde{c})T(\tilde{b}).
\end{aligned}$$

The equality $1/\det_1 T(\tilde{c})T(\tilde{b}) = \det T(a)T(a^{-1})$ can be verified by an easy computation based on 1.8(c).

Now suppose $a \in W^{\alpha,\beta}_{N\times N}$ and $\alpha + \beta > 1/2$. Then $H(b)H(\tilde{c}) \in \mathfrak{C}_2(l^2)$ due to Lemma 6.19, and similarly as above this leads to

$$\det_2(I - F_{n,o}) - \det_2(I - F_{n,o} - \sum_{k=1}^{\infty} F_{n,k}) = o(1) \quad (n \to \infty).$$

Thus, starting from 6.15(20),

$$\begin{aligned}
&\frac{G(a)^{n+1}}{D_n(a)} \exp\{tr(F_{n,o} + \sum_{k=1}^{\infty} F_{n,k})\} \\
&= \det_1(I - F_{n,o} - \sum_{k=1}^{\infty} F_{n,k}) \exp\{tr(F_{n,o} + \sum_{k=1}^{\infty} F_{n,k})\} \\
&= \det_2(I - F_{n,o} - \sum_{k=1}^{\infty} F_{n,k}) \\
&= \det_2(I - F_{n,o}) + o(1) = \det_1(I - F_{n,o}) \exp tr F_{n,o} + o(1) \\
&= \det_1(I - P_nH(\tilde{c})H(b)P_n) \exp tr P_nH(\tilde{c})H(b)P_n + o(1) \\
&= \det_2(I - P_nH(\tilde{c})H(b)P_n) + o(1) \\
&= \det_2(I - H(\tilde{c})H(b)) + o(1) = \det_2 T(\tilde{c})T(\tilde{b}) + o(1);
\end{aligned}$$

(here we used 1.9(b),(c) and the obvious identities $\det_1 W_nAW_n = \det_1 P_nAP_n$, $tr\, W_nAW_n = tr\, P_nAP_n$).

which by Horn's lemma implies the assertion. Δ

6.20. Theorem. Let $a(t)$ satisfy one of the conditions 6.18(i), (ii) and suppose $T(a)$ and $T(\tilde{a})$ are invertible on l^2_N. Define b and c by 6.15(15) and $F_{n,k}$ by 6.15(17),(18). Then if $\alpha + \beta > 1/p$, $p \in \mathbb{Z}_+$,

$$\|F_{n,k}\| = o(1/n^{(\alpha+\beta)k}) \quad \text{if 6.17(i) is satisfied;}$$

$$\|F_{n,k}\| = O(1/n^{(\alpha+\beta)k}) \quad \text{if 6.17(ii) is satisfied;}$$

$$T(\tilde{c})T(b) - I \in \mathfrak{C}_p(l^2_N);$$

$$\lim_{n\to\infty} \frac{D_n(a)}{G(a)^{n+1}} \exp\{-\sum_{j=1}^{p-1} \operatorname{tr} \frac{1}{j}(\sum_{k=0}^{p-2} F_{n,k})^j\}$$

$$= 1/\det_p T(\tilde{b})T(\tilde{c}) \quad \text{if } p \geqq 2;$$

$$\lim_{n\to\infty} D_n(a)/G(a)^{n+1} = 1/\det_1 T(\tilde{b})T(\tilde{c}) = E(a) \quad \text{if } p = 1.$$

Here $\det_p$ denotes p-regularized determinant defined in 1.9.

Remark. On comparing the Theorems 6.18 and 6.20 we see that the auxiliary terms $F_{n,k}$ in 6.20 are of somewhat more intricate structure than those in 6.18 defined via the $G_{n,k}$'s. On the other hand, 6.20 gives an information about the term $\tilde{E}(a)$. But contrary to 6.18 the Theorem 6.20 gives only $o(1)$ for the remainder in the asymptotic formulas.

Proof. To begin with, note that

$$T(\tilde{c})T(\tilde{b}) - I = -H(\tilde{c})H(b) \in \mathfrak{C}_p(l^2_N)$$

by virtue of Lemma 6.19.
At first we consider the case where $a \in W^{\alpha,\beta}_{N\times N}$ and $\alpha + \beta > 1$. Our aim is to remove the sum $\sum_{k=1}^{\infty} F_{n,k}$ in 6.15(20). According to 1.8(a) we have

$$|\det_1(I - F_{n,0}) - \det_1(I - F_{n,0} - \sum_{k=1}^{\infty} F_{n,k})|$$

$$\leqq \|\sum_{k=1}^{\infty} F_{n,k}\|_1 \exp(2\|F_{n,0}\|_1 + \|\sum_{k=1}^{\infty} F_{n,k}\|_1 + 1).$$

Since $H(b)H(\tilde{c}) \in \mathfrak{C}_1(l^2_N)$ by Lemma 6.19, we get

$$\|F_{n,k}\|_1 \leqq \|c\|_{L^\infty} \|Q_n H(b)H(\tilde{c})Q_n\|_1^k \|b\|_{L^\infty}$$

This will give the assertion once we prove that

$$\exp \operatorname{tr} \sum_{k=1}^{\infty} F_{n,k} \to 1 \quad (n \to \infty).$$

But if $k \geqq 1$ then

$$|\operatorname{tr} F_{n,k}| = |\operatorname{tr} P_n T(c)(Q_n H(b)H(\tilde{c})Q_n)^k T(b)P_n|$$

$$\leqq \sum_{j=0}^{n-1} |\operatorname{tr}(P_{j+1}-P_j)T(c)(Q_n H(b)H(\tilde{c})Q_n)^k T(b)(P_{j+1}-P_j)|$$

$$\leqq \sum_{j=0}^{n-1} \|(P_{j+1}-P_j)T(c)(Q_n H(b)H(\tilde{c})Q_n)^k T(b)(P_{j+1}-P_j)\|$$

$$\leqq \sum_{j=0}^{n-1} \|(P_{j+1}-P_j)T(c)Q_n\| \cdot \|Q_n H(b)\|^k \cdot \|H(\tilde{c})Q_n\|^k \cdot \|Q_n T(b)(P_{j+1}-P_j)\|$$

$$= O\Big(\sum_{j=0}^{n-1} (n-j)^{-\alpha} \cdot n^{-\beta k} \cdot n^{-\alpha k} \cdot (n-j)^{-\beta}\Big) = O(n^{1-(\alpha+\beta)(k+1)})$$

and therefore

$$|\operatorname{tr} \sum_{k=1}^{\infty} F_{n,k}| = O(n^{1-2(\alpha+\beta)}) = o(1) \quad (n \to \infty).$$

Now it is clear how to proceed for $a \in W_{N\times N}^{\alpha,\beta}$ if $\alpha + \beta > 1/3$, $\alpha + \beta > 1/4$ and so on.
Finally, in the case where (ii) holds the same arguments with only minor modifications apply. Δ

<u>Example.</u> Take $a \in W_{N\times N}^{\alpha,\beta}$, $\alpha + \beta > 1$, and let both $T(a)$ and $T(\tilde{a})$ be invertible on l_N^2. Then Theorems 6.20, 6.18, and Remark 3 in 6.18 give

$$\log D_n(a) = (n+1)\log G(a) + \log E(a) - \operatorname{tr}\sum_{k=1}^{\infty} k\, c_{-n-k} b_{n+k} +$$

$$+ o(1/n^{2(\alpha+\beta)-1}) \quad (n \to \infty),$$

where

$$\operatorname{tr} \sum_{k=1}^{\infty} k\, c_{-n-k} b_{n+k} = o(1/n^{\alpha+\beta-1}) \quad (n \to \infty).$$

<u>6.21. Lemma.</u> Let $a \in L_{N\times N}^{\infty}$ and assume $a_n = 0$ for all $n > n_o \geqq 0$. If $T(a)$ is Fredholm on l_N^2 then $a \in G(C + \overline{H^{\infty}})_{N\times N}$. Moreover, if both $T(a)$ and $T(\tilde{a})$ are invertible on l_N^2, then $a(t)$ admits a factorization $a = u_- u_+ = v_+ v_-$, where $u_-, v_- \in G\overline{H_{N\times N}^{\infty}}$, $u_+, v_+ \in GH_{N\times N}^{\infty}$ and u_+, v_+ are polynomials (in t) of degree at most n_o.

<u>Proof.</u> Clearly, $a \in (C + \overline{H^{\infty}})_{N\times N}$. If $T(a) \in \Phi(l_N^2)$ then $a \in G(C+\overline{H_{N\times N}^{\infty}})$

by Theorem 2.30. Moreover, we know $a \in C^\gamma + \overline{H^\infty}$ (with an arbitrary $\gamma > 0$) and so Proposition 4.23 gives $a \in G(C^\gamma + H^\infty)_{N\times N}$. By 5.23(h) the algebra $C^\gamma + H^\infty$ has the factorization property. Thus if $T(a)$ and $T(\tilde{a})$ are invertible on l^2_N then $a = u_- u_+ = v_+ v_-$, where $u_-, v_- \in G\overline{H^\infty_{N\times N}}$ and $u_+, v_+ \in GH^\infty_{N\times N}$.
Finally, from $u_+ = P(u_-^{-1} a)$ and $v_+ = P(a v_-^{-1})$ it follows that u_+ and v_+ are polynomials of degree at most n_o. Δ

6.22. Theorem. Let $a \in L^\infty_{N\times N}$ and assume $a_n = 0$ for all $n > n_o \geqq 0$ or for all $n < -n_o \leqq 0$. Suppose $T(a)$ and $T(\tilde{a})$ are invertible on l^2_N. Then there is a real number $q > 1$ such that as $n \to \infty$

$$\log D_n(a) = (n+1)\log G(a) + \log E(a) + O(1/q^n).$$

Comment. By Lemma 6.21 we have $a \in G(C + H^\infty)_{N\times N}$ or $a \in G(C + \overline{H^\infty})_{N\times N}$, so $G(a)$ is well-defined. Furthermore, the invertibility of both $T(a)$ and $T(\tilde{a})$ is by virtue of Proposition 2.29 equivalent to that of both $T(a)$ and $T(a^{-1})$. Thus $T(a)T(a^{-1}) - I \in \mathfrak{C}_1(l^2_N)$ since

$$T(a)T(a^{-1}) - I = - H(a)H(\widetilde{a^{-1}}) = - T^{-1}(a)H(a^{-1})H(\tilde{a})T(a^{-1})$$

and at least one of the operators $H(a)$ or $H(\tilde{a})$ has finite rank. Therefore, $E(a)$ is well-defined, too.

Proof. We restrict ourselves to the case where $a_n = 0$ for all $n > n_o \geqq 0$. From Lemma 6.21 it follows that 6.15(a),(b) are satisfied and consequently, the arguments of 6.15, 6.16 apply. From the theorem in 1.18 we conclude that all zeros of $\det u_+$ ly outside the closed unit disk, that is,

$$\min\{|z| : \det u_+(z) = 0\} =: \alpha > 1.$$

Thus u_+^{-1} is analytic in an open disk with radius α and we have a power series representation

$$u_+^{-1}(z) = \sum_{n\geqq 0} w_n z^n \quad (|z| < \alpha).$$

Now let q be an arbitrary real number such that $1 < q < \alpha$. By Cauchy's inequality (cf. Rudin [1], 10.25)

$$|w_n| \leqq \frac{M(q)}{q^n}, \quad \text{where } M(q) = \max\{|u_+^{-1}(z)| : |z| = q\}.$$

This gives

$$\|Q_n T(u_+^{-1}) P_o\| = \|Q_n T(u_+^{-1} - P_n u_+^{-1}) P_o\|$$

$$\leqq \|u_+^{-1} - P_n u_+^{-1}\|_W = \sum_{k=n+1}^{\infty} |w_n| \leqq \frac{M(q)}{q-1} q^{-n},$$

$$\|Q_nH(u_+^{-1})\| = \|Q_nH(u_+^{-1} - P_nu_+^{-1})\|$$
$$\leq \|u_+^{-1} - P_nu_+^{-1}\|_W \leq M(q)(q-1)^{-1}q^{-n} ,$$

and as in the proof of Theorem 6.18 we obtain

$$\|G_{n,k}\| = O(q^{-(k+1)n}) \quad (n \to \infty).$$

The computations in 6.16 lead to

$$\log D_n(a) = (n+1)\log G(a) + \log \widetilde{E}(a) + O(q^{-n})$$

and the corollary in 6.8 gives $\widetilde{E}(a) = E(a)$. Δ

6.23. The methods developed in 6.15 – 6.20 enable us to study a variety of other classes of generating functions. We refer in this connection to Silbermann [2], [3], Böttcher, Silbermann [1] and restrict ourselves here to give only an example.

Define for $0 \leq \gamma < 1/2$

$$\mathcal{L}^\gamma = \{a \in W: PM(a)P,\ QM(a)P,\ QM(a)Q \in \mathcal{B}(l_2^{\gamma,0})$$
$$\text{and } PM(a)Q \in \mathfrak{C}_2(l_2^{\gamma,0})\}$$
(cf. 4.13). Note that by 1.5

$$PM(a)Q \in \mathfrak{C}_2(l_2^{\gamma,0}) \iff Pa \in l^{\frac{1}{2}-\gamma} \quad (0 \leq \gamma < 1/2)$$

and that the algebra $W^{\gamma,0} \cap l_2^{0,1/2-\gamma}$ (see 5.23(b),(c)) is contained in $\mathcal{L}^\gamma$. Moreover, $\mathcal{L}^\gamma$ itself is easily seen to be an algebra and it is not difficult to deduce from Theorem 5.21 that $\mathcal{L}^\gamma$ has the factorization property. Finally, a straightforward computation shows that from $a \in \mathcal{L}^\gamma_{N\times N}$ and the invertibility of $T(a)$ and $T(\tilde{a})$ on l_N^2 it follows that $a^{-1} \in \mathcal{L}^\gamma_{N\times N}$. Thus if $a \in \mathcal{L}^\gamma_{N\times N}$, $T(a) \in G\mathcal{B}(l_N^2)$ and $T(\tilde{a}) \in G\mathcal{B}(l_N^2)$, then $a(t)$ admits a canonical factorization $a = u_-u_+ = v_+v_-$ in $\mathcal{L}^\gamma_{N\times N}$ and the arguments of 6.15 apply. We define b and c by 6.15(15) and have $H(b)H(\tilde{c}) \in \mathfrak{C}_2(l_N^2)$ since

$$PM(b)Q \cdot QM(c)P \in \mathfrak{C}_2(l_2^{\gamma,0}).$$

In view of $H(\tilde{c})H(b) = (H(b^\tau)H(\tilde{c}^\tau))^\tau$, we get $H(\tilde{c})H(b) \in \mathfrak{C}_2(l_N^2)$, too. Now we are in position to conclude as in 6.20 that

$$\lim_{n\to\infty} \frac{G(a)^{n+1}}{D_n(a)} \exp \operatorname{tr} \sum_{k=0}^{\infty} F_{n,k} = \det{}_2 T(\tilde{c})T(\tilde{b}).$$

We assert that $\exp \operatorname{tr} \sum_{k=2}^{\infty} F_{n,k} \to 1$ as $n \to \infty$.

Indeed, this follows from

$$|\operatorname{tr} \sum_{k=2}^{\infty} F_{n,k}| \leq \| \sum_{k=2}^{\infty} F_{n,k} \|_1$$

$$\leq \|c\|_{L^\infty} \| (Q_n H(b) H(\tilde{c}) Q_n)^2 \|_1 \left(\sum_{k=2}^{\infty} \| Q_n H(b) H(\tilde{c}) Q_n \|_\infty^{k-2} \right) \|b\|_{L^\infty}$$

and

$$\| (Q_n H(b) H(\tilde{c}) Q_n)^2 \|_1 \leq \| Q_n H(b) H(\tilde{c}) Q_n \|_2^2 = o(1) \quad (n \to \infty).$$

Thus what we have proved is:

If $a \in \mathcal{L}^{\gamma}_{N\times N}$ and if $T(a)$ and $T(\tilde{a})$ are invertible on l^2_N then

$$\lim_{n\to\infty} \frac{D_n(a)}{G(a)^{n+1}} \exp\{-\operatorname{tr}(F_{n,0} + F_{n,1})\} = 1/\det_2 T(\tilde{c}) T(\tilde{b}).$$

Note that by virtue of Theorem 6.20 the term $F_{n,1}$ may be removed if

$$a \in W^{\gamma+\varepsilon, 1/2-\gamma}_{N\times N} (\subset (W^{\gamma,0} \cap l_2^{0,1/2-\gamma})_{N\times N} \subset \mathcal{L}^{\gamma}_{N\times N}).$$

NONVANISHING INDEX

6.24. Theorem. Let $a \in C^{\alpha}_{N\times N}$, $\alpha > 1/2$, and suppose $T(a)$ and $T(\tilde{a})$ are invertible on l^2_N. Thus we have a canonical factorization $a = u_- u_+ = v_+ v_-$ and we put as usual $b = v_- u_+^{-1}$ and $c = u_-^{-1} v_+$. Then for every positive integer $\varkappa$

$$D_n \left[\begin{pmatrix} t^{-\varkappa} & & \\ & \ddots & \\ & & t^{-\varkappa} \end{pmatrix} a(t) \right] = (-1)^{(n+\varkappa)\varkappa N} G(a)^{n+1} E(a) G(c)^{\varkappa} \times$$

$$\times \left\{ \det \begin{pmatrix} b_{n+1} & \cdots & b_{n-\varkappa+2} \\ \vdots & \ddots & \vdots \\ b_{n+\varkappa} & \cdots & b_{n+1} \end{pmatrix} + O(n^{-3\alpha}) \right\} \{1 + O(n^{1-2\alpha})\}$$

and

$$D_n \left[\begin{pmatrix} t^{\varkappa} & & \\ & \ddots & \\ & & t^{\varkappa} \end{pmatrix} a(t) \right] = (-1)^{(n+\varkappa)\varkappa N} G(a)^{n+1} E(a) G(b)^{\varkappa} \times$$

$$\times \left\{ \det \begin{pmatrix} c_{-n-1} & \cdots & c_{-n+\varkappa-2} \\ \vdots & \ddots & \vdots \\ c_{-n-\varkappa} & \cdots & c_{-n-1} \end{pmatrix} + O(n^{-3\alpha}) \right\} \{1 + O(n^{1-2\alpha})\}.$$

Proof. We remember of the following fact (see Gantmacher [1], p.20): if $A = (a_{ij})_{i,j=1}^m$ is a regular matrix and $B = A^{-1}$, then

$$B\binom{i_1 \dots i_p}{k_1 \dots k_p} = \frac{(-1)^{\sum_{r=1}^{p}(i_r+k_r)} A\binom{k'_1 \dots k'_{m-p}}{i'_1 \dots i'_{m-p}}}{A\binom{1 \dots m}{1 \dots m}}, \tag{24}$$

where $C\binom{i_1 \dots i_s}{k_1 \dots k_s}$ $(i_1 < \dots < i_s,\ k_1 < \dots < k_s)$ denotes the determinant of that submatrix of C which is formed by the rows $i_1, \dots, i_s$ and the columns $k_1, \dots, k_s$, and where $\{j'_1, \dots, j'_{m-s}\} = \{1, \dots, m\} \setminus \{j_1, \dots, j_s\}$.

Now set $A = T_n(a)$ and apply (24) to the $\varkappa N \times \varkappa N$-submatrix X standing at the left lowest corner of $B = T_n^{-1}(a)$ and consisting of $\varkappa \times \varkappa$ $(N \times N)$-blocks. It results

$$\det X = \frac{(-1)^{n\varkappa N} D_{n-\varkappa}[(\begin{smallmatrix} & & t^{-\varkappa} \\ & \cdot^{\cdot^{\cdot}} & \\ t^{-\varkappa} & & \end{smallmatrix})\, a(t)]}{D_n(a)}. \tag{25}$$

Next, X may be identified with $(P_n - P_{n-\varkappa})T_n^{-1}(a)P_{\varkappa-1}$, application of formula 3.8 gives

$$\begin{aligned}
&(P_n - P_{n-\varkappa})T_n^{-1}(a)P_{\varkappa-1} \\
&= (P_n - P_{n-\varkappa})T^{-1}(a)P_{\varkappa-1} \\
&\quad - (P_n - P_{n-\varkappa})T^{-1}(a)Q_n(Q_nT^{-1}(a)Q_n)^{-1}Q_nT^{-1}(a)P_{\varkappa-1}
\end{aligned}$$

and after an appropriate computation (we refer to Böttcher, Silbermann [1] for details) we arrive at

$$\begin{aligned}
&(P_n - P_{n-\varkappa})T_n^{-1}(a)P_{\varkappa-1} \\
&= (P_n - P_{n-\varkappa})T(v_-^{-1})(P_n - P_{n-\varkappa}) \cdot (P_n - P_{n-\varkappa})T(b)P_{\varkappa-1} \times \\
&\quad \times P_{\varkappa-1}T(u_-^{-1})P_{\varkappa-1} \\
&\quad - (P_n - P_{n-\varkappa})T(u_+^{-1})T(c)Q_n \sum_{k=0}^{\infty} (Q_nH(b)H(\tilde{c})Q_n)^k \times \\
&\quad \times Q_nT(b)P_{\varkappa-1}T(u_-^{-1})P_{\varkappa-1}.
\end{aligned} \tag{26}$$

Estimates similar to those in 6.18 yield

$$\begin{aligned}
&(P_n - P_{n-\varkappa})T_n^{-1}(a)P_{\varkappa-1} \\
&= (P_n - P_{n-\varkappa})T(v_-^{-1})(P_n - P_{n-\varkappa}) \cdot (P_n - P_{n-\varkappa})T(b)P_{\varkappa-1} \times \\
&\times P_{\varkappa-1}T(u_-^{-1})P_{\varkappa-1} + O(n^{-3\alpha}) \quad (n \to \infty).
\end{aligned} \tag{27}$$

Now take the determinant of both sides of (27) and make use of the following elementary fact:

If $\{A_n\}_{n=1}^{\infty}$ and $\{B_n\}_{n=1}^{\infty}$ are sequences of mxm-matrices such that $\sup_n\|A_n\| < \infty$ and $\|A_n - B_n\| = O(\alpha_n)$ $(n \to \infty)$ then

$$\det A_n - \det B_n = O(\alpha_n) \quad (n \to \infty).$$

Hence

$$\det(P_n-P_{n-\varkappa})T_n^{-1}(a)P_{\varkappa-1}$$
$$= \det T_{\varkappa-1}(v_-^{-1})\cdot\det(P_n-P_{n-\varkappa})T(b)P_{\varkappa-1}\cdot\det T_{\varkappa-1}(u_-^{-1}) + O(n^{-3\alpha})$$
$$= G(v_-^{-1})^{\varkappa}G(u_-^{-1})^{\varkappa}\det(P_n-P_{n-\varkappa})T(b)P_{\varkappa-1} + O(n^{-3\alpha})$$

and putting this, (25), Theorem 6.18, and the identity $G(a)/G(v_-)G(u_-) = G(c)$ together we obtain the first part of the assertion. The second can be proved analogously. Δ

SEMIRATIONAL GENERATING FUNCTIONS

<u>6.25. Definition.</u> We call a matrix function $a \in GL^{\infty}_{N\times N}$ <u>semirational</u> if the matrix Fourier coefficients of a^{-1} vanish for all sufficiently large or for all sufficiently small indices, that is ,

$$(a^{-1})_n = 0 \quad \text{for all} \quad n > n_o \geqq 0 \quad \text{or for all} \quad n < -n_o \leqq 0. \qquad (28)$$

<u>6.26. Lemma.</u> Let $a \in GL^{\infty}_{N\times N}$ be semirational such that $(a^{-1})_n = 0$ for all $n < -n_o \leq 0$. Then $a \in G(C + H^{\infty})_{N\times N}$ if only $T(a) \in \Phi(l^2_N)$. Moreover, if both $T(a)$ and $T(\tilde{a})$ are invertible on l^2_N, then we have a factorization $a = u_-u_+ = v_+v_-$, where $u_-, v_- \in G\overline{H^{\infty}_{N\times N}}$, $u_+, v_+ \in GH^{\infty}_{N\times N}$ and u_-^{-1}, v_-^{-1} are matrix polynomials (in $1/t$) of degree at most n_o.

<u>Proof.</u> From Remark 3 in 2.34 we know that for $a \in GL^{\infty}_{N\times N}$

$$T(a) \in \Phi(l^2_N) \iff T(\widetilde{a}^{-1}) \in \Phi(l^2_N). \qquad (29)$$

Now our hypotheses imply $\widetilde{a}^{-1} \in (C + \overline{H^{\infty}})_{N\times N}$, and combining (29) with Theorem 2.35 we get $a \in G(C + H^{\infty})_{N\times N}$. To complete the proof it remains to repeat the arguments of the proof of Lemma 6.21. Δ

<u>6.27. Theorem.</u> Suppose $a \in GL^{\infty}_{N\times N}$ is a semirational matrix function satisfying 6.25(28). If $T(a) \in \Phi(l^2_N)$ and $\operatorname{Ind} T(a) = 0$ then there is an $n_1 \geqq 0$ such that

$$D_n(a) = G(a)^{n+1}E(a) \quad \text{for all} \quad n \geqq n_1.$$

Comment. Recall the comment following Theorem 6.22. Note thereby that in accordance with 6.9 the term $E(a)$ has to be set 0 if one of the operators $T(a)$ or $T(a^{-1})$ is not invertible.

Proof. It suffices to consider only the case where $(a^{-1})_n = 0$ for $n < -n_o \leqq 0$. Then $a \in G(C + H^\infty)_{N\times N}$.

Step 1. Suppose both $T(a)$ and $T(a^{-1})$ are invertible. Then, due to Lemma 6.26, the assumptions of 6.15 are satisfied and, moreover,

$$H(\widetilde{u_-^{-1}})Q_n = 0, \quad H(\tilde{c})Q_n = 0, \quad P_oT(c)Q_nT(b)P_o = 0$$

for all $n \geqq n_o$. Thus 6.15(19) holds for $n \geqq n_o$ and since $G_{n,k} = 0$ for $n \geqq n_o$, it takes the form $D_{n-1}(a)/D_n(a) = 1/G(a)$, whence

$$D_n(a)/G(a)^{n+1} = \tilde{E}(a) = \text{const.}$$

for all $n \geqq n_o-1$. Finally, $\tilde{E}(a) = E(a)$ by the corollary in 6.8. Notice that identity 6.17 can also be applied to prove this step.

Step 2. Assume $T(a^{-1})$ is not invertible, but let $T(a)$ be invertible. By Theorem 2.35 we have $T(a^{-1}) \in \Phi(l^2_N)$ and $\operatorname{Ind} T(a^{-1}) = 0$. So in view of Theorem 6.5 there can be found a polynomial matrix ψ such that $T(a^{-1} + \varepsilon\psi) \in G\mathcal{B}(l^2_N)$ for all ε with $0 < |\varepsilon| < r_o$. Note that $a^{-1} + \varepsilon\psi \in G(C + H^\infty)_{N\times N}$. We put $a_\varepsilon = (a^{-1} + \varepsilon\psi)^{-1}$ and have

$$\|T(a_\varepsilon) - T(a)\| = \|a_\varepsilon - a\| \leqq \|a\| \frac{\|\varepsilon a\psi\|}{1-\|\varepsilon a\psi\|} \to 0 \quad (\varepsilon \to 0).$$

Consequently, there is an $r_1 \leqq r_o$ such that $T(a_\varepsilon) \in G\mathcal{B}(l^2_N)$ for all ε with $0 < |\varepsilon| < r_1$. Then the arguments of step 1 apply and we get

$$D_n(a_\varepsilon) = G(a_\varepsilon)^{n+1} \det T(a_\varepsilon)T(a_\varepsilon^{-1}) \tag{30}$$

for $n \geqq n_1 = \max\{n_o, \deg Q\psi\}$ and $0 < |\varepsilon| < r_1$. Since all functions in (30) are analytic with respect to ε, equality holds for $\varepsilon = 0$, too. Thus $D_n(a) = 0$ for all $n \geqq n_1$.

Step 3. If $T(a) \notin G\mathcal{B}(l^2_N)$ but $T(a^{-1}) \in G\mathcal{B}(l^2_N)$, then $T(\tilde{a}) \in G\mathcal{B}(l^2_N)$, $T(\widetilde{a^{-1}}) \notin G\mathcal{B}(l^2_N)$ and $(\widetilde{a^{-1}})_n = 0$ for $n > n_o \geqq 0$. Thus step 2 gives $D_n(\tilde{a}) = 0$ for $n \geqq n_1$ and therefore

$$D_n(a) = \det T_n(a) = \det W_nT_n(\tilde{a})W_n = D_n(\tilde{a}) = 0 \quad \text{for } n \geqq n_1.$$

Step 4. Finally, suppose neither $T(a)$ nor $T(\tilde{a})$ is invertible. Since $T(a^{-1}) \in \Phi(l^2_N)$ and $\operatorname{Ind} T(a^{-1}) = 0$, there is a matrix polynomial ψ such that $T(a^{-1}+ \varepsilon\psi) \in G\mathcal{B}(l^2_N)$ in $0 < |\varepsilon| < r_1$. Put $a_\varepsilon = (a^{-1}+ \varepsilon\psi)^{-1}$ and fix ε_o such that $0 < |\varepsilon_o| < r_1$. Then

$$T(a_{\varepsilon_0}^{-1}) \in G\mathcal{B}(l_N^2), \quad T(\tilde{a}_{\varepsilon_0}) \in G\mathcal{B}(l_N^2). \tag{31}$$

Since $T(\widetilde{a_{\varepsilon_0}^{-1}})$ is Fredholm of index zero, there is a matrix polynomial φ such that

$$T[\widetilde{a_{\varepsilon_0}^{-1}} + \delta\varphi] \in G\mathcal{B}(l_N^2) \quad \text{in} \quad 0 < |\delta| < r_2. \tag{32}$$

Because of (31) there is an r_3 such that

$$T[(\widetilde{a_{\varepsilon_0}^{-1}} + \delta\varphi)^{-1}] \in G\mathcal{B}(l_N^2) \quad \text{on} \quad |\delta| = r_3 < r_2. \tag{33}$$

Now (33) and (32) give

$$T[\widetilde{a_{\varepsilon}^{-1}} + \delta\varphi] \in G\mathcal{B}(l_N^2), \quad T[(\widetilde{a_{\varepsilon}^{-1}} + \delta\varphi)^{-1}] \in G\mathcal{B}(l_N^2) \tag{34}$$

for $|\delta| = r_3$ and $|\varepsilon - \varepsilon_0| < r_4$ with some r_4. Thereby

$$(\widetilde{a_{\varepsilon}^{-1}} + \delta\varphi)_n = (\widetilde{a^{-1}} + \varepsilon\tilde{\psi} + \delta\varphi)_n = 0 \tag{35}$$

for $n < -n_1 = -\max\{n_0, \deg Q\tilde{\psi}, \deg Q\varphi\}$. Thus (34), (35) and step 1 yield

$$D_n[(\widetilde{a_{\varepsilon}^{-1}} + \delta\varphi)^{-1}] = G[(\widetilde{a_{\varepsilon}^{-1}} + \delta\varphi)^{-1}]^{n+1}E[(\widetilde{a_{\varepsilon}^{-1}} + \delta\varphi)] \tag{36}$$

for $n \geqq n_1$, $|\delta| = r_3 < r_2$ and $|\varepsilon - \varepsilon_0| < r_4$. Since both sides of (36) are analytic in $|\delta| < r_2$, we obtain equality at $\delta = 0$, thus

$$D_n(\tilde{a}_{\varepsilon}) = G(\tilde{a}_{\varepsilon})^{n+1}E(a_{\varepsilon}) \tag{37}$$

for $n \geqq n_1$ and $|\varepsilon - \varepsilon_0| < r_4$. The functions occuring in (37) are analytic in $|\varepsilon| < r_1$ and, consequently, (37) holds at $\varepsilon = 0$. This gives $D_n(\tilde{a}) = G(\tilde{a})^{n+1}E(\tilde{a})$ and from

$$E(\tilde{a}) = \det T(\tilde{a})T(\widetilde{a^{-1}}) = 0$$

we conclude $D_n(a) = 0$ for all $n \geqq n_1$, which completes the proof. Δ

<u>Remark.</u> From the proof it is seen that under the hypotheses of Theorem 6.27

$$D_n(a) = G(a)^{n+1}E(a) \neq 0 \quad \text{for all} \quad n \geqq \max\{n_0 - 1, 0\}$$

if both $T(a)$ and $T(\tilde{a})$ are invertible, and that

$$D_n(a) = 0 \quad \text{for all} \quad n \geqq n_1$$

otherwise. We do not know whether $n_0 = n_1$ or not.

<u>6.28. Theorem.</u> Assume $a \in GL_{N\times N}^{\infty}$ is a semirational matrix function such that $(a^{-1})_n = 0$ for all $n < -n_0 \leqq 0$ and suppose $T(a)$

and $T(\tilde{a})$ are invertible on l^2_N. Let $a = u_-u_+ = v_+v_-$ be the factorization given by Lemma 6.26 and set $b = v_-u_+^{-1}$, $c = u_-^{-1}v_+$. Then for every positive integer $\varkappa$ and for all $n \geq n_0$

$$D_n\left[\begin{pmatrix} t^{-\varkappa} & & \\ & \ddots & \\ & & t^{-\varkappa}\end{pmatrix} a(t)\right] =$$

$$= (-1)^{(n+\varkappa)nN} G(a)^{n+1} E(a) G(c)^{\varkappa} \det \begin{pmatrix} b_{n+1} & \cdots & b_{n-\varkappa+2} \\ \vdots & \ddots & \vdots \\ b_{n+\varkappa} & \cdots & b_{n+1}\end{pmatrix}$$

and

$$D_n\left[\begin{pmatrix} t^{\varkappa} & & \\ & \ddots & \\ & & t^{\varkappa}\end{pmatrix} a(t)\right] = 0.$$

Analogous formulas hold in the case where $(a^{-1})_n = 0$ for all $n > n_0 \geq 0$. Then one has only to replace b_k by c_{-k}, $G(c)$ by $G(b)$ and $\varkappa$ by $-\varkappa$.

<u>Proof.</u> Since $H(\tilde{c})Q_n = 0$ for $n \geq n_0$ we see that $\sum_{k=1}^{\infty} = 0$ in 6.24(26). But the 0 th item vanishes also, since v_-^{-1} is a polynomial and therefore

$$(P_n - P_{n-\varkappa})T(u_+^{-1})T(c)Q_n$$

$$= (P_n - P_{n-\varkappa})T(u_+^{-1}c)Q_n - (P_n - P_{n-\varkappa})H(u_+^{-1})H(\tilde{c})Q_n$$

$$= (P_n - P_{n-\varkappa})T(u_+^{-1}c)Q_n = (P_n - P_{n-\varkappa})T(v_-^{-1})Q_n = 0.$$

Now the first formula in our theorem follows as in 6.24.

For $P_{\varkappa-1}T_n^{-1}(a)(P_n - P_{n-\varkappa})$ an expansion may be derived which is similar to that obtained above for $(P_n - P_{n-\varkappa})T_n^{-1}(a)P_{\varkappa-1}$, and this would lead to our second formula. Instead of doing so, let us consider 6.24(26), but suppose $(a^{-1})_n = 0$ for all $n > n_0 \geq 0$. Then $b_n = 0$ for $n > n_0$ and therefore $Q_nT(b)P_{\varkappa-1} = 0$ for $n \geq n_0 + \varkappa$. Hence 6.24(25),(26) look now as follows:

$$\det X = \det(P_n - P_{n-\varkappa})T(v_-^{-1})(P_n - P_{n-\varkappa}) \times$$

$$\times \det(P_n - P_{n-\varkappa})T(b)P_{\varkappa-1} \cdot \det P_{\varkappa-1}T(u_-^{-1})P_{\varkappa-1}$$

for $n \geq n_0 + \varkappa$. But $\det(P_n - P_{n-\varkappa})T(b)P_{\varkappa-1} = 0$ for $n \geq n_0 + \varkappa$, hence

$$D_n\left[\begin{pmatrix} t^{-\varkappa} & & \\ & \ddots & \\ & & t^{-\varkappa}\end{pmatrix} a(t)\right] = 0 \qquad \text{for } n \geq n_0$$

and this is the analogue of the second formula. Δ

6.29. Theorem (K.M. Day). Let the rational function f be given by

$$f(t) = \frac{c_0 \prod_{j=1}^{p} (t-r_j)}{\prod_{j=1}^{h} (1-\frac{t}{\varrho_j}) \prod_{j=1}^{k} (t-\delta_j)} \qquad (|t| = 1), \tag{38}$$

where $|\delta_j| < 1$ $(j=1,\ldots,k)$, $|\varrho_j| > 1$ $(j=1,\ldots,h)$, c_0 is some constant, and assume $r_1,\ldots,r_p$ are pairwise distinct complex numbers. Then for $n \geqq k$

$$D_n(f) = 0 \qquad \text{if } p < k \tag{39}$$

and

$$D_n(f) = (-1)^{(p-k)(n+1)} \sum_M A_M r_M^{n+1} \qquad \text{if } p \geqq k, \tag{40}$$

where the sum is taken over all $\binom{p}{k}$ subsets $M \subset \{1,\ldots,p\}$ of cardinality k, and with $\overline{M} = \{1,\ldots,p\} \setminus M$ the A_M's and r_M's are given as

$$r_M = c_0 \prod_{j\in\overline{M}} r_j, \tag{41}$$

$$A_M = \prod_{\substack{j\in\overline{M}\\ \alpha\in\{1,\ldots,k\}}} (r_j-\delta_\alpha) \prod_{\substack{i\in M\\ \beta\in\{1,\ldots,h\}}} (\varrho_\beta-r_i) \times \tag{42}$$

$$\times \prod_{\substack{\alpha\in\{1,\ldots,k\}\\ \beta\in\{1,\ldots,h\}}} (\varrho_\beta-\delta_\alpha)^{-1} \prod_{\substack{i\in M\\ j\in\overline{M}}} (r_j-r_i)^{-1}.$$

Remark. Note that every rational function $f \in L^\infty$, which has no multiple zeros on Γ, admits the representation (38). If f has multiple zeros on Γ then $D_n(f)$ can be evaluated from (39) - (42) by means of an appropriate passage to a limit (e.g., if $r_1 = r_2$ then write down (40) for $r_1 \neq r_2$ and then take the limit $\lim_{r_1\to r_2} \ldots$ in (40)).

Proof. We shall derive our theorem from Theorem 6.28. To that end notice that it suffices to give a proof only for the case where $|r_j| > 1$ $(j = 1,\ldots,p)$. Indeed, $D_n(f)$ as well as the right hand sides of (39) and (40) are analytic functions in r_i if all $r_j \neq r_i$ are fixed and therefore, if (39),(40) holds in

$$\{r_i \in \mathbb{C} : |r_i| > 1, \quad r_i \neq r_j \ (j \neq i)\}$$

then (39),(40) will be true also in

$$\{r_i \in \mathbb{C} : r_i \neq r_j \ (j \neq i)\}.$$

Thus let $|r_j| > 1$ $(j = 1,\dots,p)$ and write (38) as

$$f(t) = \frac{c_0 \prod_{j=1}^{p}(-r_j) \prod_{j=1}^{p}(1 - \frac{t}{r_j})}{\prod_{j=1}^{k} t \prod_{j=1}^{h}(1- \frac{t}{\varrho_j}) \prod_{j=1}^{k}(1- \frac{\delta_j}{t})} = t^{-k} c_0 \prod_{j=1}^{p}(-r_j) a(t),$$

where $a = a_- a_+$ and

$$a_-(t) = \prod_{j=1}^{k}(1 - \frac{\delta_j}{t})^{-1}, \quad a_+(t) = \prod_{j=1}^{p}(1 - \frac{t}{r_j}) \prod_{j=1}^{h}(1 - \frac{t}{\varrho_j})^{-1}.$$

Obviously, $a = a_- a_+$ is a canonical Wiener-Hopf factorization and $(a^{-1})_n = 0$ for all $n < -k \leq 0$. From 6.3 we get $G(a) = 1$, $G(c) = G(a_-^{-1} a_+) = 1$ and Theorem 6.10 gives

$$E(a) = \prod_{\substack{\alpha \in \{1,\dots,k\} \\ \gamma \in \{1,\dots,p\}}} (1 - \frac{\delta_\alpha}{r_j}) \prod_{\substack{\alpha \in \{1,\dots,k\} \\ \beta \in \{1,\dots,h\}}} (1 - \frac{\delta_\alpha}{\varrho_\beta}).$$

Next, we need the Fourier coefficients b_{n+s} $(s = 1,2,\dots)$ of $b = a_- a_+^{-1}$. We have

$$b(t) = \frac{a_-(t)}{a_+(t)} = \prod_{j=1}^{h}(1 - \frac{t}{\varrho_j}) \prod_{j=1}^{k}(1 - \frac{\delta_j}{t})^{-1} \prod_{j=1}^{p}(1 - \frac{t}{r_j})^{-1}$$

and a straightforward but somewhat extensive computation leads to

$$b_{n+s} = \sum_{j=1}^{p} B_j \, 1/r_j^{n+s},$$

$$B_j = \prod_{i \neq j}(1 - \frac{r_j}{r_i})^{-1} \prod_{l=1}^{h}(1 - \frac{r_j}{\varrho_l}) \prod_{l=1}^{k}(1 - \frac{\delta_l}{r_j})^{-1} .$$

Thus

$$\det \begin{pmatrix} b_{n+1} & \cdots & b_{n-k+2} \\ \vdots & \ddots & \vdots \\ b_{n+k} & \cdots & b_{n+1} \end{pmatrix}$$

$$= \det \begin{pmatrix} \sum B_j r_j^{-(n+1)} & \cdots & \sum B_j r_j^{-(n-k+2)} \\ \vdots & \ddots & \vdots \\ \sum B_j r_j^{-(n+k)} & \cdots & \sum B_j r_j^{-(n+1)} \end{pmatrix} \quad (\sum = \sum_{j=1}^{p})$$

$$= \det \begin{pmatrix} B_1 & \cdots & B_p \\ B_1 r_1^{-1} & \cdots & B_p r_p^{-1} \\ \vdots & & \vdots \\ B_1 r_1^{-(k-1)} & \cdots & B_p r_p^{-(k-1)} \end{pmatrix} \begin{pmatrix} r_1^{-(n+1)}, & r_1^{-n} & \cdots & r_1^{-(n-k+2)} \\ \vdots & \vdots & & \vdots \\ r_p^{-(n+1)}, & r_p^{-n} & \cdots & r_p^{-(n-k+2)} \end{pmatrix}$$

which is zero for $p < k$, whereas it for $p \geqq k$ by the Cauchy-Binet formula becomes

$$= \sum_{\substack{i_1,\ldots,i_k \\ 1 \leqq i_1 < \ldots < i_k \leqq p}} \det \begin{pmatrix} B_{i_1} & \cdots & B_{i_k} \\ \vdots & & \vdots \\ B_{i_1} r_{i_1}^{-(k-1)} & \cdots & B_{i_k} r_{i_k}^{-(k-1)} \end{pmatrix} \times$$

$$\times \det \begin{pmatrix} r_{i_1}^{-(n+1)} & \cdots & r_{i_1}^{-(n-k+2)} \\ \vdots & & \vdots \\ r_{i_k}^{-(n+1)} & \cdots & r_{i_k}^{-(n-k+2)} \end{pmatrix}$$

$$= \sum \frac{B_{i_1} \cdots B_{i_k}}{(r_{i_1} \cdots r_{i_k})^{n+1}} \det \begin{pmatrix} 1 & \cdots & 1 \\ \vdots & & \vdots \\ r_{i_1}^{-(k-1)} & \cdots & r_{i_k}^{-(k-1)} \end{pmatrix} \det \begin{pmatrix} 1 & \cdots & r_{i_1}^{k-1} \\ \vdots & & \vdots \\ 1 & \cdots & r_{i_k}^{k-1} \end{pmatrix}$$

$$= \sum \frac{B_{i_1} \cdots B_{i_k}}{(r_{i_1} \cdots r_{i_k})^{n+1}} \prod_{s>t} \left(\frac{1}{r_{i_s}} - \frac{1}{r_{i_t}}\right) \prod_{s>t} (r_{i_s} - r_{i_t})$$

$$= \sum_M \frac{\prod_{i \in M} B_i}{(\prod_{i \in M} r_i)^{n+1}} \prod_{\substack{i \in M, j \in M \\ i \neq j}} (r_i - r_j) \prod_{i \in M} r_i^{-(k-1)} .$$

On substituting these things into the first formula in Theorem 6.28 we get after some computation the assertion. Δ

As a consequence of Theorem 6.27 we obtain the following result.

<u>6.30. Theorem (I.I. Hirschman, Jr.).</u> Suppose $a \in L^\infty$, $D_n(a) \neq 0$, and the polynomials p_n and q_n defined by

$$p_n = T_n^{-1}(a)\chi_0, \quad q_n = T_n^{-1}(\bar{a})\chi_0 \qquad (43)$$

have no zeros on the closed disk $|z| \leqq 1$. Then $D_{n-1}(a) \neq 0$ and

$$D_n(a) = \left(\frac{D_n(a)}{D_{n-1}(a)}\right)^{n+1} \exp\{-\frac{1}{2\pi i}\int_{-\pi}^{\pi} \log p_n(e^{i\varphi}) d \log \overline{q_n(e^{i\varphi})}\}. \tag{44}$$

Proof. Suppose $D_n(a) \neq 0$. We use the following fact, which may be found in Gohberg, Feldman [1], ch.III, 6.3:

The sequence $\{a_j\}_{j=-n}^{n}$ has an extension to a sequence $\{b_j\}_{j=-\infty}^{\infty} \in \tilde{1}^1$, $b_j = a_j$ $(-n \leqq j \leqq n)$, such that

$$b(t) = \sum_{j=-\infty}^{\infty} b_j t^j \neq 0 \quad (|t| = 1), \quad \text{ind } b = 0,$$

$$(b^{-1})_k = 0 \quad \text{for all} \quad |k| > n,$$

if and only if $p_n(z) \neq 0$ and $q_n(z) \neq 0$ for $|z| \leqq 1$. In that case

$$b(t) = (q_n)_o \overline{q_n^{-1}(t)}\, p_n^{-1}(t) \quad (|t| = 1). \tag{45}$$

Actually we need only the (almost trivial) sufficiency part of this result. Thus define b by (45). Then Theorem 6.27 and the remark in 6.27 give

$$D_k(b) = G(b)^{k+1} E(b) \quad \text{for} \quad k \geqq n-1.$$

Obviously, $D_k(b) = D_k(a)$ for $k \leqq n$. Furthermore, by Cramer's rule $D_{n-1}(a)/D_n(a) = (p_n)_o = p_n(0) \neq 0$ and therefore $D_{n-1}(a) \neq 0$. Consequently,

$$G(b) = D_n(b)/D_{n-1}(b) = D_n(a)/D_{n-1}(a)$$

and to show (44) we are in view of Theorem 6.10 left with the verification of the equality

$$\sum_{k=1}^{\infty} k(\log b)_k(\log b)_{-k} = -\frac{1}{2\pi i}\int_{-\pi}^{\pi} \log p_n(e^{i\varphi}) d \log \overline{q_n(e^{i\varphi})}.$$

But this follows by an easy computation from (45). Δ

6.31. We define $1_{p,q}^{\alpha,\beta}$ as the Banach space of all sequences $\varphi = \{\varphi_j\}_{j\in\mathbb{Z}}$ satisfying

$$\|\varphi\| = \left(\sum_{j=-\infty}^{-1} |\varphi_j|^p (|j|+1)^{p\alpha}\right)^{1/p} + \left(\sum_{j=0}^{\infty} |\varphi_j|^q (j+1)^{q\beta}\right)^{1/q} < \infty,$$

identify functions with their sequence of Fourier coefficients, and consider the Banach space $W \cap 1_{p,q}^{\alpha,\beta}$. Under certain conditions, $W \cap 1_{p,q}^{\alpha,\beta}$ can be shown to be actually a Banach algebra. For instance, this is true if

(a) $p = q = 2$, $\alpha \geqq 0$, $\beta \geqq 0$ (see 5.23(c)) or if

(b) $p > 1$, $q > 1$, $1/p + 1/q = 1$, $\alpha > 1/q$, $\beta > 1/p$ (see 4.13(b)).

But once we know that $W \cap l^{\alpha,\beta}_{p,q}$ forms an algebra, it is easy to see that it has the factorization property and that $a \in W \cap l^{\alpha,\beta}_{p,q}$ is invertible in $W \cap l^{\alpha,\beta}_{p,q}$ if and only if $a(t) \neq 0$ ($|t| = 1$). Moreover, in this case for every $a \in G(W \cap l^{\alpha,\beta}_{p,q})$ of index zero the finite section method is applicable to both $T(a)$ and $T(\bar{a})$ on l^1, l^{α}_p and l^{β}_q (cf. 4.13(12)). Let us prove this for $T(a)$ on l^{β}_q. With $a = a_-a_+$ being the Wiener-Hopf factorization we have

$$T(a) = T(a_+)T(a_-) + H(a_+)H(\tilde{a}_-),$$

because of

$$\|Q_nH(a_+)\|_{\mathcal{B}(l^{\beta}_q)} \leqq \|Q_na_+\|_{W\cap l^{\alpha,\beta}_{p,q}} = o(1) \quad (n \to \infty)$$

we get $H(a_+)H(\tilde{a}_-) \in \mathfrak{C}_{\infty}(l^{\beta}_q)$ by 1.1 and therefore it remains to show that $T(a_+)T(a_-) \in \Pi_{l^{\beta}_q}\{P_n\}$ (cf. Gohberg, Feldman [1], ch.II, §3). But this is trivial due to

$$(P_nT(a_+)T(a_-)P_n)^{-1}P_n = T_n(a_-^{-1})T_n(a_+^{-1})$$

and Proposition 3.3.

It follows an application of Theorem 6.30.

<u>6.32. Theorem.</u> Let $p > 1$, $q > 1$, $1/p + 1/q = 1$, $\alpha \geqq 0$, $\beta \geqq 0$, $\alpha + \beta \geqq 1$, and suppose $W \cap l^{\alpha,\beta}_{p,q}$ forms an algebra. Then for $a \in W \cap l^{\alpha,\beta}_{p,q}$, $a(t) \neq 0$ ($|t| = 1$), $\operatorname{ind} a = 0$, we have

$$\lim_{n\to\infty} D_n(a)/G(a)^{n+1} = E(a).$$

<u>Proof.</u> Since the finite section method is applicable to $T(a)$ and $T(\overline{a})$, we get

$$p_n = T_n^{-1}(a)\chi_0 \to T^{-1}(a)\chi_0 = (a_-^{-1})_0 \cdot a_+^{-1}, \tag{46}$$

$$q_n = T_n^{-1}(\bar{a})\chi_0 \to T^{-1}(\bar{a})\chi_0 = (\bar{a}_+^{-1})_0 \cdot \bar{a}_-^{-1} \tag{47}$$

where $a = a_-a_+$ is the Wiener-Hopf factorization and $\to$ denotes convergence in the norm of $W \cap l^{\alpha,\beta}_{p,q}$. Thus the hypotheses of Theorem 6.30 are fulfilled for all sufficiently large n, whence

$$D_n(a) = \left(\frac{D_n(a)}{D_{n-1}(a)}\right)^{n+1}E_n(a),$$

where $E_n(a) = \exp \sum_{k=1}^{\infty} k(\log p_n)_k(\log \overline{q}_n)_{-k}$. Now recall formula 6.15(19). Since $H(b)$ and $H(\tilde{c})$ are in $\mathfrak{C}_\infty(l^q)$, we obtain $\|Q_n H(b)H(\tilde{c})Q_n\|_{\mathcal{B}(l^q)} \to 0$ as $n \to \infty$. Furthermore,

$$\|Q_n T(b)P_o\|_{\mathcal{B}(l^q)} = (\sum_{k=1}^{\infty} |b_{n+k}|^q)^{1/q} = o(1/n^\beta),$$

$$\|P_o T(c)Q_n\|_{\mathcal{B}(l^q)} = (\sum_{k=1}^{\infty} |c_{-n-k}|^p)^{1/p} = o(1/n^\alpha).$$

Thus formula 6.15(19) gives as $n \to \infty$

$$\left(\frac{D_n(a)}{D_{n-1}(a)}\right)^{n+1} = G(a)^{n+1}(1 + o(\frac{1}{n^{\alpha+\beta}}))^{n+1} = G(a)^{n+1}(1 + o(1)).$$

Finally, from (46), (47) we conclude

$$\|\log p_n - \log (a_-^{-1})_o + \log a_+\|_{W\cap l^{\alpha,\beta}_{p,q}} \to 0$$

$$\|\log \overline{q}_n - \log (a_+^{-1})_o + \log a_-\|_{W\cap l^{\alpha,\beta}_{p,q}} \to 0$$

as $n \to \infty$, which implies that $E_n(a) \to E(a)$ $(n \to \infty)$ and this completes the proof. Δ

SELF-ADJOINT GENERATING FUNCTIONS

6.33. Let $a \in L^\infty_{N\times N}$ be a matrix function such that $M(a) \in \mathcal{B}(L^2_N)$ is positive definite, that is,

$$(M(a)\varphi,\varphi) = (\varphi,M(a)\varphi) \geqq m\|\varphi\|^2_{L^2_N} \quad \text{for every } \varphi \in L^2_N \qquad (48)$$

with some constant $m > 0$. Clearly, from (48) it follows that $a \in GL^\infty_{N\times N}$. Furthermore, from (48) we can deduce that $a(t)$ admits the following factorization (see Clancey, Gohberg [2], ch.VIII, Prop.4.2):

$$a = u_-u_+ = v_+v_-, \qquad (49)$$

where $u_+,v_+ \in GH^2_{N\times N}$ and $u_-,v_- \in G\overline{H^2_{N\times N}}$;

$$u_-^* = u_+, \quad v_-^* = v_+. \qquad (50)$$

Consideration of the diagonal entries of $u_+^*u_+ = a$ shows that actually $u_+ \in L^\infty_{N\times N}$; analogously $u_+^{-1} \in L^\infty_{N\times N}$ can be shown and so on. Thus

$$u_+,v_+ \in GH^\infty_{N\times N} \text{ and } u_-,v_- \in G\overline{H^\infty_{N\times N}}, \qquad (51)$$

which implies the boundedness of $T(u_+^{-1})$, $T(u_-^{-1})$ etc. on H^2_N and therefore,

$$T(a),\ T(\tilde{a}),\ T(a^{-1}) \in G\mathcal{B}(H^2_N). \qquad (52)$$

6.34. Theorem. Let $a \in L^\infty_{N\times N}$ and suppose $M(a)$ is positive definite on L^2_N. Then $T(a) \in \Pi_2\{P_n\}$.

Proof. There are several ways to show this theorem. We choose one based on 6.15. If $M(a)$ is positive definite then so are all operators $T_n(a) : \operatorname{Im} P_n \to \operatorname{Im} P_n$, consequently, all they are invertible. In view of 6.15(16) the uniform boundedness of $\|T_n^{-1}(a)P_n\|$ will follow, once we have shown that $\|Q_n H(b)H(\tilde{c})Q_n\| < 1$ holds. To that end notice the following fact (see Rabindranathan [1] or Clancey, Gohberg [2], ch.VIII, Lemma 5.1):

Let u be a unitary valued element of $L^\infty_{N\times N}$. In order that $T(u)$ be left invertible on $\mathcal{L}(H^2_N)$ it is necessary and sufficient that

$$\operatorname{dist}_{L^\infty_{N\times N}} (u, H^\infty_{N\times N}) < 1$$

(for the scalar case see also Theorem 2.15). But 6.33(49) implies that $\tilde{b} = \tilde{v}_- \tilde{u}_+^{-1}$ and $c = u_-^{-1} v_+$ are unitary valued, so 6.33(49),(51) give the invertibility of $T(\tilde{b})$ and $T(c)$. Hence there are h and k in $H^\infty_{N\times N}$ such that $\|c - h\|_{L^\infty_{N\times N}} < 1$, $\|\tilde{b} - k\|_{L^\infty_{N\times N}} < 1$, and we obtain

$$\|Q_n H(b)H(\tilde{c})Q_n\| \leqq \|H(b)H(\tilde{c})\| = \|H(b-\tilde{k})H(\tilde{c}-\tilde{h})\| \leqq$$

$$\leqq \|b-\tilde{k}\|\|\tilde{c}-\tilde{h}\| = \|\tilde{b}-k\|\|c-h\| < 1. \quad \Delta$$

Remark. Thus, although condition 6.15(b) is, in general, not satisfied for positive definite generating functions, the proof of Theorem 6.34 shows that the arguments of 6.15 apply.

6.35. Lemma. Let $a \in L^\infty_{N\times N}$ and suppose $M(a)$ is positive definite on L^2_N. Define $G_{n,k}$ $(n,k \geqq 0)$ by 6.15(13)-(15). Then the operators $G_{n,k} : \operatorname{Im} P_0 \to \operatorname{Im} P_0$ are positive, i.e.

$$(G_{n,k}\varphi,\varphi) = (\varphi,G_{n,k}\varphi) \geqq 0 \text{ for every } \varphi \in \operatorname{Im} P_0 , \qquad (53)$$

and

$$\Big\|\sum_{k=0}^{\infty} G_{n,k}\Big\|_{\mathcal{L}(\operatorname{Im} P_0)} = o(1) \quad \text{as } n \to \infty. \qquad (54)$$

Proof. We have $(b_n)^* = c_{-n}$, where b_n and c_n are the matrix Fourier coefficients of $b = v_- u_+^{-1}$ and $c = u_-^{-1} v_+$, respectively. Thus $T(b)^* = T(c)$ and $H(b)^* = H(\tilde{c})$, whence (53) follows by a straightforward computation.
To prove (54) note that by 6.15(12)

$$P_0 - \sum_{k=0}^{\infty} G_{n,k} = T_0(u_+)P_0 T_n^{-1}(a)P_0 T_-(u_-),$$

that by virtue of Theorem 6.34 and 1.3(d)

$$\|P_0 T_n^{-1}(a) P_0 - P_0 T^{-1}(a) P_0\| \to 0 \quad \text{as} \quad n \to \infty$$

and therefore,

$$\sum_{k=0}^{\infty} G_{n,k} = P_0 - T_0(u_+) P_0 T^{-1}(a) P_0 T_0(u_-) + o(1)$$

$$= P_0 - T_0(u_+) T_0(u_+^{-1}) T_0(u_-^{-1}) T_0(u_-) + o(1) = o(1). \quad \Delta$$

<u>6.36. Theorem.</u> Let $a \in L^{\infty}_{N\times N}$ and suppose $M(a)$ is positive definite on L^2_N. Assume there holds for infinitely many $n \geqq n_0$

$$D_n(a) = \mathcal{G}(a)^{n+1} \mathcal{E}(a) \tag{55}$$

with any constants $\mathcal{G}(a)$ and $\mathcal{E}(a)$. Then there is an n_1 such that $(a^{-1})_n = 0$ for $|n| \geqq n_1$. In the scalar case one can take $n_1 = n_0 + 1$. Moreover, in the scalar case the following is true: if there exists an n_0 such that

$$D_{n_0}(a)/D_{n_0-1}(a) = \exp(\log a)_0, \tag{56}$$

then $(a^{-1})_n = 0$ for $|n| \geqq n_0 + 1$.

<u>Proof.</u> First of all, $\mathcal{G}(a) \neq 0$ and $\mathcal{E}(a) \neq 0$ since $T(a) \in \Pi_2\{P_n\}$ by Theorem 6.34. Next, if (55) holds for infinitely many n then by virtue of the corollary in 6.1 and 6.33(49),(51)

$$\mathcal{G}(a) = \det P_0 T^{-1}(a) P_0,$$

whence because of 6.15(19)

$$\det(I - \sum_{k=0}^{\infty} G_{n,k}) = \mathcal{G}(a) \frac{D_{n-1}(a)}{D_n(a)} = 1 \tag{57}$$

for infinitely many $n \geqq n_0$. Furthermore, if (56) is satisfied then (57) holds also for $n = n_0$. Now let $\lambda_1^{(n)}, \ldots, \lambda_N^{(n)}$ denote the eigenvalues of the $N \times N$-matrix $\sum_{k=0}^{\infty} G_{n,k}$. Then (57) may be rewritten as

$$\prod_{i=1}^{N} (1 - \lambda_i^{(n)}) = 1. \tag{58}$$

From Lemma 6.35 we know that $\lambda_i^{(n)} \geqq 0$ and that $\lambda_i^{(n)} \to 0$ as $n \to \infty$. Thus if $N = 1$ then (58) gives $\lambda_1^{(n_1)} = 0$ for $n_1 = n_0$ and if $N > 1$ then (58) yields $\lambda_i^{(n)} = 0$ for a sufficiently large n, which will be also denoted by n_1. Hence $\sum_{k=0}^{\infty} G_{n,k} = 0$, which because of 6.35(53) implies that $G_{n_1,0} = 0$. From Remark 3 in 6.18 we get therefore

$$\sum_{k=1}^{\infty} c_{-n_1-k} b_{n_1+k} = \sum_{k=1}^{\infty} b^*_{n_1+k}\, b_{n_1+k} = 0,$$

i.e. $b^*_n b_n = 0$, and thus $b_n = 0$ and $c_{-n} = b^*_n = 0$ for all $n \geqq n_1+1$. So u_+^{-1} and u_-^{-1} are polynomials (in t and $1/t$, respectively) of degree at most n_1 and consequently, $(a^{-1})_n = 0$ for $n \geqq n_1+1$. Δ

6.37. The Theorems 6.11(6.14), 6.20, 6.22 and 6.32 give rise to the following observation: in order to ensure the validity of the Szegö limit theorem, the smoothness of Pa can be weakened if, simultanously, the smoothness condition on Qa is strengthened. For a self-adjoint generating function, however, both Pa and Qa have equal smoothness. In this "symmetrical" case the minimal smoothness sufficient for the Szegö limit theorem to hold is given by the following theorem.

6.38. Theorem. Let $a \in L^{\infty}_{N\times N}$ and suppose $M(a)$ is positive definite on L^2_N. Then $\det P_0 T^{-1}(a) P_0 \neq 0$ and we define $G(a)$ as

$$G(a) = 1/\det P_0 T^{-1}(a) P_0$$

(for $a \in (C + H^{\infty})_{N\times N}$ this agress with 6.3). Now, if the finite limit

$$\lim_{n\to\infty} D_n(a)/G(a)^{n+1} = A \tag{59}$$

exists, then $a \in (K^{1/2,1/2}_{2,2})_{N\times N}$.

Proof. Since $a(t)$ admits the factorization 6.33(49)-(51), we have

$$\det P_0 T^{-1}(a) P_0 = \det T_0(u_+^{-1}) \cdot \det T_0(u_-^{-1}) =$$

$$= 1/\det[u_+(0) u_+^*(0)] \in (0,\infty),$$

hence $G(a) \in (0,\infty)$ (that $\det P_0 T^{-1}(a) P_0$ does not vanish is also a consequence of Lemma 6.12). Furthermore, the positive definiteness of $T(a)$ gives that $D_n(a) \in (0,\infty)$ for all n.
Theorem 6.34 shows that formula 6.15(19) applies. Put $g_n = \sum_{k=0}^{\infty} G_{n,k}$. From Lemma 6.35 it follows that $\det(P_0 - g_n) \in (0,1)$ for all n large enough. Thus

$$D_{n-1}(a)/D_n(a) = \det(P_0 - g_n)/G(a) \leqq 1/G(a)$$

and therefore,

$$0 < D_n(a)/G(a)^{n+1} < D_{n+1}(a)/G(a)^{n+2},$$

which implies that $A \neq 0$. Now for sufficiently large n and m, $n < m$,

$$\prod_{i=n+1}^{m} \det(P_o - g_i) = \prod_{i=n+1}^{m} \frac{D_{i-1}(a)}{D_i(a)} G(a) =$$

$$= \frac{D_n(a)}{G(a)^{n+1}} \cdot \frac{G(a)^{m+1}}{D_m(a)} \to \frac{D_n(a)}{G(a)^{n+1}} \cdot \frac{1}{A} \quad (m \to \infty),$$

whence we obtain the convergence of the infinite product $\prod_{i=n+1}^{\infty} \det(P_o - g_i)$. Since $\|g_i\| < 1$ for sufficiently large i due to Lemma 6.35, we get so the convergence of the series

$$\sum_{i=n+1}^{\infty} \log \det(P_o - g_i) = \sum_{i=n+1}^{\infty} \operatorname{tr} \log(P_o - g_i)$$

and again by Lemma 6.35,

$$- \log(P_o - g_i) = g_i + g_i^2/2 + g_i^3/3 + \dots \geqq$$

$$\geqq g_i = \sum_{k=o}^{\infty} G_{i,k} \geqq G_{i,o} \geqq 0.$$

Consequently the series $\sum_{i=n+1}^{\infty} \operatorname{tr} G_{i,o}$ is convergent. But by Remark 3 in 6.18 and $c_{-k} = b_k^*$, we have

$$\sum_{i=n+1}^{\infty} \operatorname{tr} G_{i,o} = \sum_{j=1}^{\infty} j \operatorname{tr}(c_{-n-j-1} b_{n+j+1}) = \sum_{j=1}^{\infty} j \operatorname{tr}(b_{n+j+1}^* b_{n+j+1})$$

and consideration of the diagonal entries of $b_{n+j+1}^* b_{n+j+1}$ gives $Pb \in (l_2^{1/2})_{N\times N}$. Therefore $H(b) \in \mathfrak{C}_2(l_N^2)$ and from

$$H(u_+) = - T(u_+)T(v_-^{-1})H(b)T(\tilde{u}_+)$$

it follows that $H(u_+) \in \mathfrak{C}_2(l_N^2)$, that is, $u_+ \in (l_2^{1/2})_{N\times N}$. Combined with 6.33(51) this gives $u_+ \in (K_{2,2}^{1/2,1/2})_{N\times N}$. From $c_{-n} = b_n^*$ we get analogously $u_- \in (K_{2,2}^{1/2,1/2})_{N\times N}$ and then $a = u_- u_+ \in (K_{2,2}^{1/2,1/2})_{N\times N}$. Δ

<u>6.39.</u> The techniques developed here allow to describe the asymptotic behavior of $D_n(a)$ for real valued functions in $K_{2,2}^{1/2,1/2}$ satisfying $\operatorname{ess\,inf} a > 0$ as follows:

$$\log D_n(a) = (n+1)(\log a)_o + \sum_{k=1}^{\infty} k(\log a)_k(\log a)_{-k} +$$

$$- \beta_n + O(\beta_n(\beta_n + \gamma_n + \beta_n\gamma_n)) \quad (n \to \infty),$$

where

$$\beta_n = \sum_{k=1}^{\infty} k|b_{n+k+1}|^2, \quad \gamma_n = \|Q_{n+1}H(b)H(\tilde{c})Q_{n+1}\|.$$

To see this take the logarithm of the identity 6.17, expand $\log(1-g_j)$ into a power series, note that

$$|g_j| \leqq \|P_0 T(c) Q_j\|_{\mathfrak{C}_2(l^2)} \|Q_j H(b) H(\tilde{c}) Q_j\|_{\mathfrak{B}(l^2)} \|Q_j T(b) P_0\|_{\mathfrak{C}_2(l^2)}$$

$$= \sum_{l=1}^{\infty} |b_{j+l+1}|^2 \|Q_j H(b) H(\tilde{c}) Q_j\|_{\mathfrak{B}(l^2)},$$

$$\|Q_j H(b) H(\tilde{c}) Q_j\| \leqq \|Q_{n+1} H(b) H(\tilde{c}) Q_{n+1}\| \text{ for } j \geqq n+1 ,$$

and put these things together.

SEPARATION AND PIECEWISE CONTINUOUS GENERATING FUNCTIONS

6.40. Theorem. Let $a, b \in L^{\infty}_{N\times N}$, suppose

$$H(a)H(\tilde{b}) \in \mathfrak{C}_1(l^2_N),\ H(\tilde{a})H(b) \in \mathfrak{C}_1(l^2_N),\ T(a) \in \Pi_2\{P_n\},\ T(b) \in \Pi_2\{P_n\} \quad (60)$$

and assume at least one of the operators $T(ab)$ or $T(\tilde{a}\tilde{b})$ is invertible on l^2_N. Then

$$\lim_{n\to\infty} \frac{D_n(ab)}{D_n(a)D_n(b)} = F(a,b)F(\tilde{a},\tilde{b}),$$

where

$$F(a,b) := \det T^{-1}(b)T^{-1}(a)T(ab). \quad (61)$$

Remark 1. It follows from (60) that both $F(a,b)$ and $F(\tilde{a},\tilde{b})$ are well-defined.

Proof. Assume $T(\tilde{a}\tilde{b}) \in G\mathfrak{B}(l^2_N)$. Then

$$P_n T(a) T(b) P_n = T_n(a) T_n(b) + W_n H(\tilde{a}) H(b) W_n$$
$$= W_n [T_n(\tilde{a}) T_n(\tilde{b}) + P_n H(\tilde{a}) H(b) P_n] W_n$$

and consequently,

$$\det P_n T(a) T(b) P_n =$$
$$= D_n(a) D_n(b) \det\{P_n + T_n^{-1}(\tilde{b}) T_n^{-1}(\tilde{a}) P_n H(\tilde{a}) H(b) P_n\}. \quad (62)$$

But if $T(a) \in \Pi_2\{P_n\}$, then also $T(\tilde{a}) \in \Pi_2\{P_n\}$ (Proof: $T_n(\tilde{a}) = W_n T(a) W_n$, use Theorem 3.16 to get the invertibility of $T(\tilde{a})$ and then apply 3.2(b)). Thus (62) gives

$$\lim_{n\to\infty} \frac{\det P_n T(a) T(b) P_n}{D_n(a) D_n(b)} = \det\{I + T^{-1}(\tilde{b}) T^{-1}(\tilde{a}) H(\tilde{a}) H(b)\} =$$
$$= \det T^{-1}(\tilde{b}) T^{-1}(\tilde{a}) T(\tilde{a}\tilde{b}).$$

Now by virtue of Theorem 3.16, $A := T(a)T(b) \in \Pi_2\{P_n\}$. Indeed, A itself is invertible, a simple computation gives

$$\tilde{A} = \operatorname*{s-lim}_{n\to\infty} W_n T(a) T(b) W_n = T(\tilde{a}\tilde{b}),$$

which has been assumed to be invertible and finally,

$$\{P_n A P_n\}^o = \{T_n(a)T_n(b) + W_n H(\tilde{a})H(b)W_n\}^o$$
$$= \{T_n(a)\}^o\{T_n(b)\}^o$$

implies that $\{P_n A P_n\}^o \in G(\mathfrak{A}/\mathfrak{J})$. To complete the proof it remains to apply Lemma 6.8 with

$$A = T(a)T(b), \quad K = T(ab) - T(a)T(b) = H(a)H(\tilde{b});$$

what results is

$$\lim_{n\to\infty} \frac{D_n(ab)}{\det P_n T(a)T(b)P_n} = \det\{I + T^{-1}(b)T^{-1}(a)H(a)H(b)\} =$$
$$= \det T^{-1}(b)T^{-1}(a)T(ab).$$

Lastly, if $T(\tilde{a}\tilde{b}) \notin G\mathfrak{B}(l^2_N)$ but $T(ab) \in G\mathfrak{B}(l^2_N)$ then replace in the above argument a by $\tilde{a}$ and b by $\tilde{b}$. Δ

<u>Remark 2.</u> Note that in the scalar case the invertibility of $T(ab)$ and $T(\tilde{a}\tilde{b})$ automatically results from (60).

<u>Remark 3.</u> Confer 2.26 for conditions ensuring that the product of two Hankel operators is of trace class.

<u>6.41.</u> Define $\varphi_r = \varphi_{t_r}^{\beta_r}$ $(r = 1,\ldots,R)$ as in 2.28, where

(a) $t_1,\ldots,t_R$ are pairwise distinct points on Γ and $\beta_1,\ldots,\beta_R$ are complex numbers satisfying $-1/2 < \mathrm{Re}\,\beta_r < 1/2$ $(r = 1,\ldots,R)$.

Furthermore, suppose

(b) $b \in L^\infty$, and to each t_r $(r = 1,\ldots,R)$ there are a neighborhood $U_r \subset \Gamma$ and a function $b_r \in B_1^1$ (recall 2.17(b)) such that $b|U_r = b_r|U_r$.

Iterative application of Theorem 6.40, together with Proposition 2.26 and Corollary 1 in 3.18, gives the following result.

<u>Corollary.</u> Let the conditions (a) and (b) be satisfied and suppose $T(b) \in \Pi_2\{P_n\}$. Then

$$\lim_{n\to\infty} \frac{D_n(\varphi_1\cdots\varphi_R b)}{D_n(\varphi_1)\cdots D_n(\varphi_R)D_n(b)} =$$

$$= F(\varphi_1\cdots\varphi_R, b)F(\tilde{\varphi}_1\cdots\tilde{\varphi}_R, \tilde{b}) \prod_{r=1}^{R-1} F(\varphi_r,\psi_r)F(\tilde{\varphi}_r,\tilde{\psi}_r),$$

where $F(\cdot,\cdot)$ is given by (61) and ψ_r is defined to be $\psi_r = \varphi_{r+1}\cdots\varphi_R$ $(r = 1,\ldots,R-1)$.

Our next concern is the determination of the asymptotic behavior of

$D_n(\varphi)$ $(n \to \infty)$, where $\varphi = \varphi_\tau^\beta$, $\tau \in \Gamma$, $-1/2 < \text{Re}\,\beta < 1/2$. Note that φ admits the representation

$$\varphi(t) = \varphi_\tau^\beta(t) = (1 - \frac{\tau}{t})^{-\beta}(1 - \frac{t}{\tau})^\beta, \quad t \in \Gamma,$$

where $(1 - t/\tau)^\beta$ denotes the limit on the unit circle of that branch of the function which is analytic in $|t| < 1$ and takes the value 1 at $t = 0$, and where $(1 - \tau/t)^{-\beta}$ is defined similarly. Put

$$\varphi_\mu(t) = (\varphi_\tau^\beta)_\mu(t) := (1 - \frac{\tau}{\mu t})^{-\beta}(1 - \frac{t}{\mu\tau})^\beta, \quad t \in \Gamma,$$

where $\mu > 1$ is a real parameter. Obviously

$$\varphi_\mu \in C^\infty(\Gamma), \quad \varphi_\mu(t) \neq 0 \quad \text{for} \quad t \in \Gamma, \quad \text{ind}\,\varphi_\mu = 0. \tag{63}$$

It is easily seen that $T(\varphi_\mu) \to T(\varphi)$ and $H(\varphi_\mu) \to H(\varphi)$ strongly as $\mu \to 1+0$. Somewhat less obvious is the following fact.

<u>6.42. Lemma.</u> $T^{-1}(\varphi_\mu) \to T^{-1}(\varphi)$ strongly as $\mu \to 1+0$.

<u>Proof.</u> From (63) we deduce the invertibility of $T(\varphi_\mu)$. It is not difficult to show that

$$T^{-1}(\varphi_\mu)e_n = T[(1 - \frac{t}{\mu\tau})^{-\beta}]T[(1 - \frac{\tau}{\mu t})^\beta]e_n$$

$$\to T[(1 - \frac{t}{\tau})^{-\beta}]T[(1 - \frac{\tau}{t})^\beta]e_n = T^{-1}(\varphi)e_n$$

with respect to the norm of l^2, where e_n is defined as in 2.4. Since the set of all finite linear combinations of $e_0, e_1, e_2, \ldots$ is dense in l^2, the assertion will follow once we have proved the uniform boundedness of the norms $\|T^{-1}(\varphi_\mu)\|_{\mathcal{B}(l^2)}$ with respect to $\mu \in (1, 1+\varepsilon)$. It is elementary function theory to show that the range of $\varphi_\mu(t)$ for $t \in \Gamma$, $\mu \in (1, 1+\varepsilon)$, is contained in a <u>closed</u> sector of an annulus spanned by the angle $2\pi|\text{Re}\,\beta| < \pi$ and having the radii $\exp\{\pm\pi|\text{Im}\,\beta|\}$. Thus, there exists a complex number $c \neq 0$ such that the disk with center in c and the radius $|c|$ contains this sector of an annulus in its interior. Therefore,

$$|\varphi_\mu(t) - c| < q|c| \quad \text{for} \quad t \in \Gamma \quad \text{and} \quad \mu \in (1, 1+\varepsilon)$$

with some q satisfying $0 < q < 1$. This implies

$$\|T(\varphi_\mu) - cI\|_{\mathcal{B}(l^2)} = \|\varphi_\mu - c\|_{L^\infty} \leqq q|c| \quad \text{for} \quad \mu \in (1, 1+\varepsilon),$$

whence by expansion into Neumann's series,

$$T^{-1}(\varphi_\mu) = \{I + \sum_{n=1}^{\infty}(-1)^n c^{-1}(T(\varphi_\mu) - cI)^n\}c^{-1},$$

and from $\|c^{-1}(T(\varphi_\mu) - cI)\|_{\mathcal{B}(l^2)} \leqq q < 1$ we get

$$\|T^{-1}(\varphi_\mu)\|_{\mathcal{B}(l^2)} \leqq c^{-1}(1-q)^{-1} \quad \text{for} \quad \mu \in (1,1+\varepsilon). \quad \Delta$$

6.43. Definition. The entire function $G(z)$ defined by

$$G(1+z) = (2\pi)^{z/2} e^{-z(z+1)/2-\gamma_E z^2/2} \times \\ \times \prod_{n=1}^{\infty} \{(1+\frac{z}{n})^n e^{-z+z^2/2n}\} \tag{64}$$

($\gamma_E = 0.577\ldots$ Euler's constant) is called the Barnes G-function. Its role in analysis becomes clear from the relation $G(1+z) = \Gamma(z)G(z)$, where $\Gamma(z)$ is the familar Gamma-function.

6.44. Theorem (Fisher-Hartwig). Let $\varphi = \varphi_\tau^\beta$, $\tau \in \Gamma$, $-1/2 < \mathrm{Re}\,\beta < 1/2$, be the function defined in 2.28. Then

$$D_n(\varphi) \sim n^{-\beta^2} G(1+\beta)G(1-\beta) \quad (n \to \infty),$$

where G denotes the Barnes G-function.

Proof. Put $\psi = \varphi_\tau^{(-\beta)}$. Then with c, $L = \mathrm{diag}(\lambda_j)_{j=0}^\infty$ and $M = \mathrm{diag}(\mu_j)_{j=0}^\infty$ as in step 1 of the proof of the theorem in 2.31, $T^{-1}(\psi) = cLT(\varphi)M$. Consequently,

$$\det\ P_n T^{-1}(\psi) P_n = c^{n+1} D_n(\varphi) \prod_{j=0}^{n} (\lambda_j \mu_j).$$

Lemma 6.42 combined with Proposition 6.7 gives

$$\det P_n T^{-1}(\psi) P_n = \lim_{\mu \to 1+0} \det P_n T^{-1}(\psi_\mu) P_n = 1.$$

An elementary computation shows

$$\prod_{j=0}^{n} (\lambda_j \mu_j) = G(1+\beta)^{-1} G(1-\beta)^{-1} \Gamma(1+\beta)^{-n-1} \Gamma(1-\beta)^{-n-1} \times \\ \times G(\beta+n+2)G(-\beta+n+2)/G(n+2)^2$$

and from $\Gamma(1+\beta)\Gamma(1-\beta) = \pi\beta/\sin\pi\beta = c$ and

$$G(n+\beta)G(n-\beta)/G(n)^2 \sim n^{+\beta^2} \quad (n \to \infty)$$

(the latter fact can be derived from 6.43(64) by a straightforward computation) we get the assertion. Δ

Combining this theorem with the corollary in 6.40 and Theorem 6.18 or Theorem 6.11 we get the following facts.

Corollary 1. Suppose the conditions 6.41(a),(b) are fulfilled. If in addition b satisfies a condition of type 6.18(i),(ii), if $H_k(b)$ $(k = 1,2,3,\ldots)$ is given as in Theorem 6.18, and if $T(b)$ is invertible on l^2, then

$$\lim_{n\to\infty} D_n(\varphi_1\dots\varphi_R b)G(b)^{-n-1}n^{\sum\beta_i^2}e^{-(H_1(b)+\dots+H_n(b))} = \tilde{E}(\varphi_1\dots\varphi_R b)$$

with some constant $\tilde{E}(\varphi_1\dots\varphi_R b) \neq 0$.

Corollary 2. Let the conditions 6.41(a),(b) be satisfied. In addition assume $b \in K^{\alpha,\beta}_{p,q}$, where $p > 1$, $q > 1$, $1/p + 1/p = 1$, $\alpha > 0$, $\beta > 0$, $\alpha + \beta = 1$, $-1/2 < \alpha - 1/p < 1/2$ $(-1/2 < \beta - 1/q < 1/2)$, and suppose $T(b)$ is invertible on l^2. Then

$$D_n(\varphi_1\dots\varphi_R b) \sim G(b)^{n+1}n^{-\sum\beta_i^2}E(b)\prod_{r=1}^{R} G(1+\beta_r)G(1-\beta_r)\times$$

$$\times F(\varphi_1\dots\varphi_R,b)F(\tilde{\varphi}_1\dots\tilde{\varphi}_R,\tilde{b})\prod_{r=1}^{R-1} F(\varphi_r,\psi_r)F(\tilde{\varphi}_r,\tilde{\psi}_r),$$

where $G(b)$ and $E(b)$ are defined as in 6.3 and 6.9, where $F(\cdot,\cdot)$ is given by (61) and $\psi_r = \varphi_{r+1}\dots\varphi_R$ $(r = 1,\dots,R-1)$.

Under the rather general hypotheses of Corollary 2, the identification of the constants $F(\varphi_1\dots\varphi_R,b)$ and $F(\tilde{\varphi}_1\dots\tilde{\varphi}_R,\tilde{b})$ is by no means trivial (and can perhaps only be based on the theory of local factorization). However, if b (globally) belongs to B_1^1, then the computation of those constants is straightforward. It rests on identity 1.10 and some approximation arguments. The following Lemmas 6.45 - 6.49 form the technical background of this computation, which will be carried out in the proof of Theorem 6.50.

Define

$$E(f,g) = \det T^{-1}(f)T^{-1}(g)T(f)T(g)$$

whenever this makes sense. In view of

$$T^{-1}(f)T^{-1}(g)T(f)T(g) = I + T^{-1}(f)T^{-1}(g)[H(g)H(\tilde{f})-H(f)H(\tilde{g})],$$

$E(f,g)$ is well-defined, if $T(f)$ and $T(g)$ are invertible on l^2 and if $H(g)H(\tilde{f})$ and $H(f)H(\tilde{g})$ are of trace class.

6.45. Lemma. Let $f,g \in L^\infty$, $c \in B_1^1$, and suppose $T(f)$, $T(g)$, $T(c)$ are invertible on l^2. If $H(f)H(\tilde{g}) \in \mathfrak{C}_1(l^2)$ then

$$E(fg,c) = E(f,c)E(g,c)$$

and if $H(\tilde{f})H(g) \in \mathfrak{C}_1(l^2)$ then

$$E(c,fg) = E(c,f)E(c,g).$$

Proof. By our hypotheses, all the terms $E(fg,c)$, $E(f,c)$,... are well-defined. Furthermore, if $H(f)H(\tilde{g}) \in \mathfrak{C}_1(l^2)$ then $T(fg)T^{-1}(g)T^{-1}(f)-I \in \mathfrak{C}_1(l^2)$. Now a few application of 1.8(c) gives

$$E(f,c) \cdot \det T(fg)T^{-1}(g)T^{-1}(f) \cdot E(g,c)$$

$$= \det T^{-1}(f)T^{-1}(c)T(f)T(c) \cdot \det T^{-1}(c)T^{-1}(f)T(fg)T^{-1}(g)T(c) \times$$

$$\times \det T^{-1}(c)T(g)T(c)T^{-1}(g)$$

$$= \det T^{-1}(f)T^{-1}(c)T(fg)T(c)T^{-1}(g)$$

and

$$E(fg,c) = \det T^{-1}(fg)T^{-1}(c)T(fg)T(c)$$

$$= \det T(f)T(g)T^{-1}(fg)T^{-1}(c)T(fg)T(c)T^{-1}(g)T^{-1}(f)$$

$$= \det T(f)T(g)T^{-1}(fg) \cdot \det T^{-1}(f)T^{-1}(c)T(fg)T(c)T^{-1}(g)$$

and since

$$\det T(f)T(g)T^{-1}(fg) \cdot \det T(fg)T^{-1}(g)T^{-1}(f) = 1,$$

we get the first assertion. The second can be proved analogously. Δ

<u>6.46. Lemma.</u> If f and c_+ are invertible of index zero in W and $W \cap H^\infty$, respectively, and if $c_+ \in B_1^1$, then

$$E(f,c_+) = \exp \sum_{k=1}^{\infty} k(\log f)_{-k}(\log c_+)_k.$$

Analogously, if g and c_- are invertible of index zero in W and $W \cap \overline{H^\infty}$, respectively, and if $c_- \in B_1^1$, then

$$E(c_-,g) = \exp \sum_{k=1}^{\infty} k(\log c_-)_{-k}(\log g)_k.$$

<u>Proof.</u> Let $f = f_-f_+$ be the canonical Wiener-Hopf factorization of f (cf. 5.23(a)). Then

$$E(f,c_+) = \det T^{-1}(f)T^{-1}(c_+)T(f)T(c_+)$$

$$= \det T(f_+^{-1})T(f_-^{-1})T(c_+^{-1})T(f_-)T(f_+)T(c_+)$$

$$= \det T(f_-^{-1})T(c_+^{-1})T(f_-)T(c_+)$$

$$= \exp\{-T(\log f_-)\} \exp\{-T(\log c_+)\} \times$$

$$\times \exp\{T(\log f_-)\} \exp\{T(\log c_+)\}$$

$$= \exp \operatorname{tr} H[\log c_+]H[(\log f_-)^{\sim}] ;$$

here we applied 1.8(c), Proposition 2.7 and identity 1.10. By Theorem 4.11 the spectrum of c_+ in W coincides with that of c_+ in $K_{1,1}^{1,1}$. Thus $\log c_+ \in K_{1,1}^{1,1}$ and therefore, $H[\log c_+] \in \mathfrak{C}_1(l^2)$.

Since

$$\operatorname{tr} H(a)H(\tilde{b}) = \sum_{k=1}^{\infty} k a_k b_{-k} \quad \text{for} \quad H(a)H(\tilde{b}) \in \mathfrak{C}_1(l^2),$$

the first assertion follows. The second can be verified similarly. Δ

<u>6.47. Lemma.</u> If $c \in C^2$ then $\|H(c)\|_1 \leqq \frac{\pi^4}{12} \|c''\|_{L^\infty}$.

<u>Proof.</u> Let p_n be the trigonometric polynomial of degree n of best uniform approximation to c. Then (cf. Korneĭčuk [1], eq. (9.5))

$$s_n(H(c)) \leqq \|H(c) - H(p_n)\|_{\mathcal{B}(l^2)} \leqq \|c - p_n\|_{L^\infty}$$

$$\leqq \frac{\pi}{2n} \omega(c', \tfrac{\pi}{n}) \leqq \frac{\pi}{2n} \|c''\|_{L^\infty} \cdot \frac{\pi}{n},$$

whence

$$\|H(c)\|_1 = \sum_{n=1}^{\infty} s_n(H(c)) \leqq \sum_{n=1}^{\infty} \frac{\pi^2}{2n^2} \|c''\|_{L^\infty} = \frac{\pi^4}{12} \|c''\|_{L^\infty}. \quad \Delta$$

<u>6.48. Lemma.</u> Let $\varphi = \varphi_\tau^\beta$, $\tau \in \Gamma$, $-1/2 < \operatorname{Re} \beta < 1/2$, be defined as in 2.28. Suppose $c \in PC$ is continuous at τ and belongs to C^2 on each closed subset of Γ containing no of its points of discontinuity. Define φ_μ as in 6.41. Then

$$\|H(\varphi_\mu)H(\tilde{c}) - H(\varphi)H(\tilde{c})\|_1 = o(1) \quad \text{as} \quad \mu \to 1+0.$$

<u>Proof.</u> Let $\sum f_i^2 = 1$ be a smooth partition of unity constructed in the same fashion as in the proof of Proposition 2.24. Then by virtue of identity 2.23

$$H(\varphi_\mu)H(\tilde{c}) - H(\varphi)H(\tilde{c}) = T[(\varphi_\mu - \varphi)c] - T(\varphi_\mu - \varphi)T(c)$$

$$= \sum \{T[(\varphi_\mu - \varphi)f_i^2 c] - T(\varphi_\mu - \varphi)T(f_i^2)T(c)\}$$

$$= \sum H[(\varphi_\mu - \varphi)f_i]H(\tilde{f}_i\tilde{c})$$

$$+ \sum T[(\varphi_\mu - \varphi)f_i]H(f_i)H(\tilde{c}) + \sum H(\varphi_\mu - \varphi)H(f_i)T(f_i c)$$

$$- \sum H(\varphi_\mu - \varphi)H(\tilde{f}_i)H(f_i)H(\tilde{c}) - \sum T(\varphi_\mu - \varphi)H(f_i)H(\tilde{f}_i)T(c).$$

But $T[(\varphi_\mu - \varphi)f_i] \to 0$, $T(\varphi_\mu - \varphi) \to 0$, and $H(\varphi_\mu - \varphi) \to 0$ strongly as $\mu \to 1+0$, and $H(f_i), H(\tilde{f}_i) \in \mathfrak{C}_1(l^2)$, so because of 1.3(d) we are left with the verification of

$$\|H[(\varphi_\mu - \varphi)f_i]H(\tilde{f}_i\tilde{c})\|_1 = o(1) \quad (\mu \to 1+0). \tag{65}$$

Among the functions f_i there is exactly one (say f_{i_0}) which is identically 1 in some neighborhood U_τ of τ. For $i = i_o$ we have $\tilde{f}_{i_o}\tilde{c} \in C^2$, hence $H[\tilde{f}_{i_o}\tilde{c}] \in \mathfrak{C}_1(l^2)$ and (65) is an immediate consequence of 1.3(d). On the other hand, if $i \neq i_o$ then $f_i|U_\tau \equiv 0$, and since for $k = 0,1,2,\dots$

$$\|\varphi_\mu^{(k)} - \varphi^{(k)}\|_{L^\infty(\Gamma \setminus U_\tau)} = o(1) \qquad (\mu \to 1+0)$$

we deduce from Lemma 6.47 that

$$\|H[(\varphi_\mu - \varphi)f_i]\|_1 \leqq \frac{\pi^4}{12} \|[(\varphi_\mu - \varphi)f_i]''\|_{L^\infty} = o(1) \qquad (\mu \to 1+0),$$

whence (69) follows at once. Δ

6.49. Lemma. If $H(b)$ and $H(\tilde{b})$ are of trace class then $b \in W$.

Proof. Let $\{e_n\}_{n=0}^\infty$ denote the standard orthonormal basis of l^2 (see 2.4). On account of 1.4 we have

$$|b_1| + |b_3| + \dots = \sum |(H(b)e_n, e_n)| < \infty,$$

$$|b_{-1}| + |b_{-3}| + \dots = \sum |(H(\tilde{b})e_n, e_n)| < \infty,$$

and in view of

$$H(\chi_1 b) = T(\chi_1)H(b) + H(\chi_1)T(\tilde{b}) \in \mathfrak{C}_1(l^2),$$

$$H(\chi_{-1}\tilde{b}) = T(\chi_{-1})\, H(\tilde{b}) + H(\chi_{-1})T(b) \in \mathfrak{C}_1(l^2),$$

we get analogously

$$|b_0| + |b_2| + \dots = \sum |(H(\chi_1 b)e_n, e_n)| < \infty,$$

$$|b_{-2}| + |b_{-4}| + \dots = \sum |(H(\chi_{-1}\tilde{b})e_n, e_n)| < \infty. \quad \Delta$$

6.50. Theorem. Suppose condition 6.41(a) is satisfied, let $b \in B_1^1$ and let $T(b)$ be invertible on l^2 (which by Lemma 6.49 is equivalent to $b(t) \neq 0$ for $t \in \Gamma$ and $\operatorname{ind} b = 0$). Then as $n \to \infty$

$$D_n(\varphi_1 \dots \varphi_R b) \sim G(b)^{n+1} n^{-(\beta_1^2 + \dots + \beta_R^2)} E(b) \prod_{r \neq s} |t_r - t_s|^{\beta_r \beta_s} \times$$
$$\times \prod_{r=1}^R G(1+\beta_r)G(1-\beta_r) \prod_{r=1}^R b_+(t_r)^{\beta_r} b_-(t_r)^{-\beta_r}.$$

Here $b_+ = \exp(P \log b)$, $b_- = \exp(Q \log b)$ are the factors from the canonical Wiener-Hopf factorization of b (ensured by Lemma 6.49), and as usual,

$$G(b) = \exp(\log b)_0, \quad E(b) = \exp \sum_{k=1}^\infty k(\log b)_{-k}(\log b)_k.$$

Proof. Because of Corollary 2 in 6.44 (note that $B_1^\ast \subset K_{2,2}^{1/2,1/2}$) we are left with the proof of the equalities

$$F(\varphi_1\ldots\varphi_R,b)F(\tilde{\varphi}_1\ldots\tilde{\varphi}_R,\tilde{b}) = \prod_{r=1}^{R} b_+(t_r)^{\beta_r} b_-(t_r)^{-\beta_r}, \qquad (66)$$

$$\prod_{r=1}^{R-1} F(\varphi_r,\psi_r)F(\tilde{\varphi}_r,\tilde{\psi}_r) = \prod_{r\neq s} |t_r - t_s|^{\beta_r\beta_s} . \qquad (67)$$

Write $\varphi = \varphi_1\ldots\varphi_R$. Then

$$\begin{aligned} F(\varphi,b) &= \det T^{-1}(b)T^{-1}(\varphi)T(b\varphi) \\ &= \det T^{-1}(b_+)T^{-1}(b_-)T^{-1}(\varphi)T(b_-)T(\varphi)T(b_+) \\ &= \det T^{-1}(b_-)T^{-1}(\varphi)T(b_-)T(\varphi) \qquad (1.8(c)) \\ &= E(b_-,\varphi) = \prod_{r=1}^{R} E(b_-,\varphi_r) \qquad (\text{Lemma } 6.45) \end{aligned}$$

(see also 5.23(e)). As a consequence of 1.3(d) and of the Lemmas 6.42 and 6.48,

$$\begin{aligned} E(b_-,\varphi_r) &= \det(I + T^{-1}(b_-)T^{-1}(\varphi_r)H(\varphi_r)H(\tilde{b}_-)) \\ &= \lim_{\mu\to1+0} \det(I + T^{-1}(b_-)T^{-1}[(\varphi_r)_\mu]H[(\varphi_r)_\mu]H(\tilde{b}_-)) \\ &= \lim_{\mu\to1+0} E(b_-,(\varphi_r)_\mu), \end{aligned}$$

and from Lemma 6.46 we deduce

$$\begin{aligned} E(b_-,(\varphi_r)_\mu) &= \exp \sum_{k=1}^{\infty} k(\log b_-)_{-k}(\log(\varphi_r)_\mu)_k \\ &= \exp \sum_{k=1}^{\infty} k(\log b_-)_{-k}(-\beta_r k^{-1}\mu^{-k}t_r^{-k}) \\ &= b_-(\mu t_r)^{-\beta_r} . \end{aligned}$$

Thus

$$F(\varphi_1\ldots\varphi_R,b) = \prod_{r=1}^{R} b_-(t_r)^{-\beta_r} .$$

Analogously can be shown that

$$F(\tilde{\varphi}_1\ldots\tilde{\varphi}_R,\tilde{b}) = \prod_{r=1}^{R} b_+(t_r)^{\beta_r} ,$$

whence (66) follows.

Now consider $F(\varphi_r,\psi_r)$. As above, we get from 1.3(d) and the Lemmas 6.42 and 6.48

$$\begin{aligned} F(\varphi_r,\psi_r) &= \det(I + T^{-1}(\psi_r)T^{-1}(\varphi_r)H(\varphi_r)H(\tilde{\psi}_r)) \\ &= \lim_{\mu\to1+0} \det(I + T^{-1}(\psi_r)T^{-1}[(\varphi_r)_\mu]H[(\varphi_r)_\mu]H(\tilde{\psi}_r)) \end{aligned}$$

$$= \lim_{\mu\to 1+0} \det T^{-1}(\psi_r)T^{-1}[(\varphi_r)_\mu]T[(\varphi_r)_\mu\psi_r]$$

$$= \lim_{\mu\to 1+0} \det T^{-1}(\psi_r)T^{-1}[(\varphi_r)_\mu^+]T(\psi_r)T[(\varphi_r)_\mu^+]$$

$$= \lim_{\mu\to 1+0} E(\psi_r,(\varphi_r)_\mu^+),$$

where $(\varphi_r)_\mu^+(t) := (1 - t/\mu t_r)^{\beta_r}$. According to Lemma 6.45,

$$E(\psi_r,(\varphi_r)_\mu^+) = \prod_{s=r+1}^{R} E(\varphi_s,(\varphi_r)_\mu^+),$$

and Lemma 6.42 gives

$$E(\varphi_s,(\varphi_r)_\mu^+) = \lim_{\lambda\to 1+0} E((\varphi_s)_\lambda,(\varphi_r)_\mu^+) .$$

Finally, on applying Lemma 6.46 we obtain

$$\begin{aligned} E((\varphi_s)_\lambda,(\varphi_r)_\mu^+) &= \exp \sum_{k=1}^{\infty} k(\log(\varphi_s)_\lambda)_{-k}(\log(\varphi_r)_\mu^+)_k \\ &= \exp \sum_{k=1}^{\infty} k(\beta_s t_s^k k^{-1}\lambda^{-k})(-\beta_r k^{-1}\mu^{-k}t_r^{-k}) \\ &= (1 - \frac{t_s}{\mu\lambda t_r})^{\beta_r\beta_s} . \end{aligned}$$

Putting these things together we arrive at

$$F(\varphi_r,\psi_r) = \prod_{s=r+1}^{R} (1 - \frac{t_s}{t_r})^{\beta_r\beta_s} .$$

Analogously one can show that

$$F(\tilde{\varphi}_r,\tilde{\psi}_r) = \prod_{s=r+1}^{R} (1 - \frac{t_r}{t_s})^{\beta_r\beta_s}.$$

Hence

$$\prod_{r=1}^{R-1} F(\varphi_r,\psi_r)F(\tilde{\varphi}_r,\tilde{\psi}_r) = \prod_{r<s} (1 - \frac{t_s}{t_r})^{\beta_r\beta_s}(1 - \frac{t_r}{t_s})^{\beta_r\beta_s}$$

$$= \prod_{r<s} |t_r - t_s|^{2\beta_r\beta_s} = \prod_{r\neq s} |t_r - t_s|^{\beta_r\beta_s}$$

and (67) is proved. Δ

6.1. - 6.6. See Widom [8].

6.7. - 6.8. The lemma in 6.8 appeared explicitly first in Böttcher [1], although it is implicitly contained also in Basor [2] and Basor, Helton [1]. The essence of the corollary in 6.8 and its proof given in the text were first published in Basor, Helton [1], but were also known to the authors, who only in view of the possibility to write this book did not make an extra publication.

6.9. The derivation of the Szegö limit theorem by means of the theory of operator determinants and thus also this definition go back to Widom [8].

6.10. This theorem is published here for the first time, although the idea to use the identity 1.10 is due to Widom [8].

6.11. This theorem represents a generalization of the case $p = q = 2$, $\alpha = \beta = 1/2$, which was considered in Widom [8] and that itself was the culmination of many earlier results of several authors.

The first results on the asymptotic behavior of Toeplitz determinants go back to Szegö. In 1952 he published the following fact: if $a \in C^{1+\varepsilon}$ $(\varepsilon > 0)$ is positive then as $n \to \infty$

$$D_n(a) \sim G(a)^{n+1}E(a), \qquad (*)$$

where $G(a) = \exp(\log a)_0$, $E(a) = \exp \sum_{k=1}^{\infty} k|(\log a)_k|^2$ (Szegö [1], see also Grenander, Szegö [1]). The smoothness condition needed by Szegö was then relaxed by a host of authors. We mention here only Kac [1] ($a \in W^{1,1}$, cf. 5.23(b) for notation), Baxter [1], [2] ($a \in W^{1/2,1/2}$), Krein [2] ($a \in C \cap l_2^{1/2,1/2}$), Devinatz [2] ($a \in L^\infty \cap l_2^{1/2,1/2}$). Moreover, Golinskiĭ, Ibragimov [1] showed then that (*) is always true if it makes sense.
For the case where the generating function is allowed to be complex-valued, and its positivity is replaced by the condition that $a(t) \neq 0$ $(t \in \Gamma)$, ind $a = 0$, we cite only Baxter [3] and Hirschman [1] ($a \in W \cap$ $\cap\, l_2^{1/2,1/2}$). Finally, Widom [8] succeeded for complex-valued generating functions in the block case and established the result mentioned above. The reader should consult 6.37 and 6.38 in order to understand in what sense the "smoothness" condition required by Widom actually represents a minimum.

6.12. - 6.13. This proof is published here for the first time.

6.14. The perturbation technique and the analyticity argument are due to Widom [8].

6.15. - 6.20. These results go back to Silbermann [2], [3] and Böttcher, Silbermann [1].

6.21. - 6.22. These results are also due to the authors and are published here for the first time.

6.23. See also Böttcher, Silbermann [1].

6.24. In this form the theorem was first proved in Böttcher, Silbermann [1]. In Fisher, Hartwig [1], [2], a similar result was stated for the scalar case, but with other methods, under different hypotheses and the higher order correction estimates were not extracted out of the determinants $\det(b_{n+1+i-j})_{i,j=1}^{\varkappa}$ and $\det(c_{-n-1+i-j})_{i,j=1}^{\varkappa}$.

6.25. - 6.28. We should emphasize here that not polynomials, but inverses of polynomials are the best possible thing in the context of Toeplitz determinants (this was perhaps already known to Szegö).

Theorem 6.27 was proved for the scalar case in Silbermann [1] and then for the block case in Böttcher, Silbermann [1]. The proof presented in text is new. For a "factorization-free" proof see Silbermann [1]. Theorem 6.28 appeared first in Böttcher, Silbermann [1].

6.29. The proof presented in the text is published for the first time. The theorem itself is due to Day [1]. For a different proof see also Høholdt, Justesen [1]. The best thing we know in this direction are the results of Gorodeckiĭ [2], who gave yet another proof and succeeded in the block case: $f(t)$ is assumed to be of the form $h^{-1}(t)k^{-1}(t)(g_{ij}(t))_{i,j=1}^{N}$, where $h(t) = \prod(1-t/\varrho_j)$, $k(t) = \prod(t-\delta_j)$ are as in Theorem 6.29, and g_{ij} are polynomials of degree at most p such that $g := \det(g_{ij})$ is a polynomial of degree pN with leading coefficient 1; then $D_n(f)$ is given for <u>all</u> $n \geqq 0$ via a determinant of constant dimension pN involving besides n only the <u>zeros</u> of g (which, moreover, are not required to be pairwise distinct!) and the <u>values</u> of h and k at the zeros of g.

6.30. - 6.32. Theorem 6.30 was proved with other methods in Hirschman [1]. The simple proof of the text is published for the first time. Confer also Baxter, Hirschman [1]. Theorem 6.32 for $p = q = 2$, $\alpha = \beta = 1/2$, goes back to Hirschman [1], too. In this more general form it was first proved in Silbermann [3].

6.36. This theorem is due to Böttcher, Silbermann [1].

6.38. In the scalar case this theorem was stated independently in Krein [2] and Devinatz [2]. A block case version was first given in Böttcher, Silbermann [1] under the hypotheses that $a \in (C + H^{\infty})_{N\times N}$.

For $L^\infty_{N\times N}$-symbols the theorem is published here for the first time. The proof of the text is adopted from Böttcher, Silbermann [1] and seems to be essentially simpler than those of Krein [2] and Devinatz [2].

6.40. This theorem (for $N = 1$) is explicitly contained in Basor [2] and implicitly also in Böttcher [1]. See also Basor [3] (however, Theorem 2 of the latter paper is not true; a counterexample is $\varphi(t) = 7 + \sum_{n=1}^{\infty} n^{-3/2}(t^{n^2}+t^{-n^2})$, $t\in\Gamma$: in fact, φ is invertible in W and has index zero, so it satisfies the hypotheses of the mentioned Theorem 2 - on the other hand, φ is positive, but $\varphi \notin K^{1/2,1/2}_{2,2}$, whence it follows by Theorem 6.38 that the Strong Szegö Limit Theorem does not hold in this case).

6.43. See Barnes [1] and Whittaker, Watson [1] for some properties of the G-function.

6.44. The theorem was first proved in Fisher, Hartwig [1]. The proof presented here is probably new.
The Corollaries 1 and 2 are published for the first time. We emphasize that b must belong to B^1_1 only in a neighborhood of the points of discontinuity of $\varphi_1 \ldots \varphi_R$.

6.50. Such a theorem was conjectured in Fisher, Hartwig [1]. In the presented form it was first proved in Basor [2]. Independently the same result was obtained in Böttcher [1], and for the case where the generating function has only one discontinuity also in Blecher [2]. The methods developed in Basor [2] and Böttcher [1] coincide, in particular, the Lemmas 6.42, 6.45, 6.46, and 6.48 are contained in both papers.

Asymptotic distribution of eigenvalues. If $a(t)$ belongs to $(K^{1/2,1/2}_{2,2})_{N\times N}$ and if both $T(a)$ and $T(a^{-1})$ are invertible on l^2_N, then by the Szegö limit theorem 6.11,

$$\log D_n(a) = (n+1)\log G(a) + \log E(a) + o(1) \quad (n \to \infty). \quad (68)$$

Denoting by $\lambda_1^{(n)},\ldots,\lambda_{(n+1)N}^{(n)}$ the eigenvalues of $T_n(a)$ we have

$$\log D_n(a) = \sum_i \log \lambda_i^{(n)}.$$

Thus, in the special case where $f(\lambda) = \log \lambda$, the Szegö limit theorem gives an answer to the question of the asymptotic behavior of the sums $\sum_i f(\lambda_i^{(n)})$ as $n \to \infty$. It is a somewhat surprising fact that, conversely, for a sufficient large class of functions $f(\lambda)$ the asymptotic behavior of such sums can almost immediately be derived from the Szegö limit theorem.

To that end note first of all that the spectrum of $a \in (K^{1/2,1/2}_{2,2})_{N\times N}$ in this algebra coincides with the spectrum of $a(t)$ in $L^\infty_{N\times N}$ (Theorem 4.11). Furthermore, if U is any open subset of $\mathbb{C}$ containing $\sigma(T(a)) \cup \sigma(T(a^{-1}))$, then U also contains the spectrum (eigenvalues) of $T_n(a)$ for all n large enough (Theorem 3.7). Since, moreover, the spectrum of $a \in L^\infty_{N\times N}$ is a subset of the spectrum of $T(a) \in \mathcal{B}(l^2_N)$ (Theorem 2.33), we conclude that $f(a) \in (K^{1/2,1/2}_{2,2})_{N\times N}$ and $f(T_n(a))$ are well-defined if only $a(t)$ belongs to $(K^{1/2,1/2}_{2,2})_{N\times N}$ and f is analytic on $\sigma(T(a)) \cup \sigma(T(a^{-1}))$. Note that in this case

$$\sum_i f(\lambda_i^{(n)}) = \operatorname{tr} f(T_n(a)). \tag{69}$$

<u>Theorem (Widom).</u> Let $a \in (C \cap K^{1/2,1/2}_{2,2})_{N\times N}$ and suppose f is analytic on the closure of some open bounded subset U of $\mathbb{C}$ containing $\sigma(T(a)) \cup \sigma(T(a^{-1}))$. Then

$$\sum_i f(\lambda_i^{(n)}) = \operatorname{tr} f(T_n(a)) = (n+1)G^f(a) + E^f(a) + o(1) \quad (n \to \infty),$$

where

$$G^f(a) = (\operatorname{tr} f(a))_0 = \frac{1}{2\pi} \int_0^{2\pi} (f(a))(e^{i\vartheta})d\vartheta,$$

$$E^f(a) = \frac{1}{2\pi i} \int_{\partial U} f(\lambda) \frac{d}{d\lambda} \log \det T[\varphi-\lambda]T[(\varphi-\lambda)^{-1}]d\lambda.$$

Remark. This theorem remains valid under the weaker hypothesis $a \in (K^{1/2,1/2}_{2,2})_{N\times N}$ (see Widom [8]).

<u>Proof.</u> (Widom [8]). If $\lambda \notin \sigma(T(a)) \cup \sigma(T(a^{-1}))$ then by (68)

$$\log \det T_n(a - \lambda) = (n+1) \log G(a - \lambda) +$$

$$+ \log \det T[a - \lambda]T[(a - \lambda)^{-1}] + o(1) \quad (n \to \infty)$$

and as is seen e.g. from 6.12 - 6.13, this holds uniformly for λ belonging to a neighborhood of ∂U. Hence, we can differentiate both sides with respect to λ, multiply by $f(\lambda)$, and integrate over ∂U. Now,

$$\int_{\partial U} f(\lambda) \frac{d}{d\lambda} \log \det T_n(a-\lambda)d\lambda = \int_{\partial U} f(\lambda) \frac{d}{d\lambda} \sum_j \log(\lambda_j^{(n)}-\lambda)d\lambda =$$

$$= \sum_j \int_{\partial U} \frac{f(\lambda)}{\lambda-\lambda_j^{(n)}} d\lambda = 2\pi i \sum_j f(\lambda_j^{(n)}),$$

and denoting by $\mu_j(e^{i\vartheta})$ the N eigenvalues of $a(e^{i\vartheta})$,

$$\int_{\partial U} f(\lambda) \frac{d}{d\lambda}(\log \det(a(e^{i\vartheta})-\lambda))_0 \, d\lambda$$

$$= (\int_{\partial U} f(\lambda) \sum_j \frac{d}{d\lambda} \log(\mu_j(e^{i\vartheta})-\lambda) \, d\lambda)_0$$

$$= (\sum_j \int_{\partial U} \frac{f(\lambda)}{\lambda-\mu_j(e^{i\vartheta})} d\lambda)_0$$

$$= 2\pi i(\sum_j f(\mu_j(e^{i\vartheta})))_0 = 2\pi i(\operatorname{tr} f(a))_0. \quad \Delta$$

Corollary. If $a \in (C \cap K_{2,2}^{1/2,1/2})_{N\times N}$ and if f is analytic on $\sigma(T(a)) \cup \sigma(T(a^{-1}))$, then

$$\lim_{n\to\infty} \frac{\sum_i f(\lambda_i^{(n)})}{(n+1)N} = \lim_{n\to\infty} \frac{\operatorname{tr} f(T_n(a))}{(n+1)N} = \frac{(\operatorname{tr} f(a))_0}{N}.$$

Proof. Immediate from the theorem. Δ

For further results along these lines we refer the reader to Grenander, Szegö [1], Hirschman, Hughes [1], Dym [1], [2], Widom [8]-[11], Guillemin [1].

Continuous analogues and higher-dimensional versions of Szegö's limit theorem.

(a) Let $k(x)$ be a matrix function defined on $\mathbb{R}^m$ such that

$$\int_{\mathbb{R}^m} \|k(x)\| dx + \int_{\mathbb{R}^m} |x| \|k(x)\|^2 dx < \infty, \quad Fk \in L^1(\mathbb{R}^m), \tag{70}$$

where $\|k(x)\|$ denotes the Hilbert-Schmidt norm of the matrix $k(x)$, and F the Fourier transform. Then $a := I_{N\times N} + Fk$ is in $L^\infty_{N\times N}(\mathbb{R}^m)$, and the (block) multiplication operator $M(a)$, defined as

$$M(a) : L^2_N(\mathbb{R}^m) \to L^2_N(\mathbb{R}^m), \quad \varphi \mapsto F^{-1}a(\cdot)F\varphi,$$

is bounded on $L^2_N(\mathbb{R}^m)$. Given a compact subset K of $\mathbb{R}^m$ with C^1 boundary denote by τK the set $\{x \in \mathbb{R}^m : x/\tau \in K\}$, by $P_{\tau K}$ the projection of $L^2_N(\mathbb{R}^m)$ onto $L^2_N(\tau K)$, that is,

$$(P_{\tau K}\varphi)(x) = \varphi(x) \text{ if } x \in \tau K, \text{ and } (P_{\tau K}\varphi)(x) = 0 \text{ if } x \notin \tau K,$$

and write $M_{\tau K}(a)$ for the operator $P_{\tau K}M(a)P_{\tau K}$ acting on $\operatorname{Im} P_{\tau K} = L^2_N(\tau K)$.

Under our hypotheses, the operator $M_{\tau K}(a)$ can be shown to differ from the identity only by a trace class operator, thus $\det M_{\tau K}(a)$ is well-defined.

<u>Theorem (Widom [12]).</u> Let Σ be the closed convex hull of the set $\bigcup (a(\xi)u,u)$, where the union is taken over all $\xi \in \mathbb{R}^m$ and $u \in \mathbb{R}^N$ such that $\|u\| = 1$. If $0 \notin \Sigma$ then there are two constants $G_K(a)$ and $E_K(a)$ such that as $n \to \infty$

$$\log \det M_{\tau K}(a) = \tau^m \log G_K(a) + \tau^{m-1} \log E_K(a) + o(\tau^{m-1}).$$

Thereby,

$$\log G_K(a) = \frac{\text{vol } K}{(2\pi)^m} \int_{\mathbb{R}^m} (\log \det a)(\xi)d\xi.$$

For an identification of $E_K(a)$ and for a proof of this result we refer to Widom [12].

(b) Let us consider the scalar case $(N = 1)$. Define $s(z) = (F^{-1}\log a)(z)$, $z \in \mathbb{R}^m$. Then, obviously,

$$\log G_K(a) = s(0) \cdot \text{vol } K.$$

The constant term $\log E_K(a)$ can be identified as

$$\log E_K(a) = \frac{1}{2} \int_{\partial K} d\sigma \int_{z\cdot\nu_\sigma \geq 0} (z\cdot\nu_\sigma)s(z)s(-z)dz, \tag{71}$$

where ν_σ is the inner unit normal to ∂K at the point σ and $d\sigma$ denotes surface measure.

(c) Now take $m = 1$ and $K = [0,1]$ (but let $N \geq 1$ be an arbitrary integer). Then the (block) Wiener-Hopf integral operator $W(a)$ on $L^2_N(\mathbb{R}_+)$ is defined in an obvious way (cf. the facts added to the notes and comments to chapter 3). If (70) is satisfied and if both $W(a)$ and $W(a^{-1})$ are invertible on $L^2_N(\mathbb{R}_+)$, then it is not too hard to see that $W(a)W(a^{-1})$ differs from the identity only by a trace class operator. The asymptotic formula takes the form

$$\log \det P_\tau W(a)P_\tau = \tau \log G(a) + \log E(a) + o(1) \quad (\tau \to \infty),$$

where $P_\tau W(a)P_\tau = M_{\tau\cdot[0,1]}(a)$,

$$\log G(a) = \log G_{[0,1]}(a) = \frac{1}{2\pi} \int_{-\infty}^{\infty} (\log \det a)(\xi)d\xi,$$

$$\log E(a) = \log E_{[0,1]}(a) = \log \det W(a)W(a^{-1}). \tag{72}$$

Apart from a few technical details this can be shown by the same methods as in the discrete case (cf. 6.7 - 6.8 and 6.12 - 6.13).

(d) Finally, choose $m = 1$, $N = 1$, $K = [0,1]$. If (70) holds, if $a(\xi) \neq 0$ for $\xi \in \mathbb{R}$ and $\text{ind } a(\xi) = 0$, then

$$\log \det P_\tau W(a)P_\tau = \tau \log G(a) + \log E(a) + o(1) \quad (\tau \to \infty), \tag{73}$$

where as above $P_\tau W(a)P_\tau = M_{\tau\cdot[0,1]}(a)$,

$$\log G(a) = s(0),\ \log E(a) = \int_0^\infty z s(z)s(-z)dz, \qquad (74)$$

$$s(z) = (F^{-1} \log a)(z).$$

The relations (73) and (74) together are the well-known formula of Achieser-Kac (cf. Achieser [1], Kac [1], Hirschman [1], Widom [6],[12]). Note that (74) can easily be derived from (71), or with the methods of 6.10 also from (72).

(e) In Widom [12] one can also find asymptotic expensions of $\operatorname{tr} f(M_{\tau K}(a))$ (cf. (69)), where f is analytic on Σ and $f(1) = 0$. In case $f(\lambda) = \log \lambda$ we obtain the theorem quoted in (a). See also Landau, Widom [1].

(f) Returning to the discrete case we mention the following result established in Linnik [1]. Let K be a bounded finitely-connected domain in $\mathbb{R}^m$ with C^2 boundary ∂K each connected component of which has positive Gaussian curvature. For a positive real number τ denote by K_τ the set of all lattice points of $\mathbb{Z}^m$ belonging to τK and by $\operatorname{card} K_\tau$ the cardinality of K_τ. Furthermore, let $a \in C(\Gamma^m)$ be a real-valued function such that $\inf a > 0$ and

$$\sum_{k\in\mathbb{Z}^m} |a_k| + \sum_{k\in\mathbb{Z}^m} |k|\,|a_k|^2 < \infty.$$

Again the m-dimensional Toeplitz operator is defined in an obvious way (cf. chapter 3), and we denote by $P_{\tau K}$ the projection of $l^2((\mathbb{Z}_+)^m)$ onto $l^2(\tau K)$. Clearly, $P_{\tau K}T^{(m)}(a)P_{\tau K}$ is of the form identity + trace class operator and has therefore a well-defined determinant. Denoting as usual by $(\log a)_k$, $k \in \mathbb{Z}^m$, the Fourier coefficients of $\log a$, we have (Linnik [1])

$$\log \det P_{\tau K}T^{(m)}(a)P_{\tau K} =$$

$$= \operatorname{card} K_\tau \cdot \log G_K(a) + \tau^{m-1} \log E_K(a) + o(\tau^{m-1}) \quad (\tau \to \infty),$$

where

$$G_K(a) = (\log a)_{(0,\dots,0)},$$

$$E_K(a) = \frac{1}{2} \sum_{k\in\mathbb{Z}^m} g(k)|(\log a)_k|^2,$$

$$g(k) = \int_{\partial K} (k \cdot \nu_\sigma)_+ d\sigma,\ (k \cdot \nu_\sigma)_+ = \frac{1}{2}(k \cdot \nu_\sigma + |k \cdot \nu_\sigma|),$$

ν_σ is the inner unit normal to ∂K at the point σ and $d\sigma$ denotes surface measure.

(g) As far as we know, in the discrete case higher-dimensional versions of Szegö's limit theorem seem to be published only for positive generating function (and thus for the scalar case). Besides the results of Linnik [1] cited in (f), we mention the paper Seghier [1], where an asymptotic formula for $\log \det(P_n \otimes P_n)T^{(2)}(a)(P_n \otimes P_n)$ (a positive) is announced. Furthermore, we do not know any reference dealing with the continuous Szegö limit theorem for regions $K \subset \mathbb{R}^m$ $(m \geq 2)$ having edges (e.g. with the asymptotic behavior of $\det(P_\tau \otimes P_\tau)W^{(2)}(a)(P_\tau \otimes P_\tau)$).

(h) Lastly, for further generalizations (manifold versions, traces of families of pseudo-differential operators etc.) we refer to Widom [9] - [11], Dym [1], [2], and Guillemin [1].

C H A P T E R 7

SINGULAR GENERATING FUNCTIONS

If a function $a(t)$ does not belong to L^∞ then $T(a)$ is unbounded on all spaces l^p, and if $a \in C \subset L^\infty$ has zeros on Γ then $T(a)$ is not Fredholm on l^2. However, there is a lot of problems leading to Toeplitz operators whose symbols have zeros or poles or other "singularities". In this connection the study of Toeplitz determinants with singular generating functions is of particular interest.
The knowledge one has of this set of problems today shows that it is hopeless to attack this question from a general point of view, but it seems reasonable to begin with the investigation of certain classes of generating functions.

THE CONJECTURE OF FISHER AND HARTWIG

7.1. The class under consideration. In Fisher, Hartwig [1], the following class of generating functions was introduced:

$$a(t) = \prod_{r=1}^{R} \left|1 - \frac{t}{t_r}\right|^{2\alpha_r} (-t)_{t_r}^{\beta_r} b(t) \quad (|t| = 1), \tag{1}$$

where

(a) b is sufficiently smooth, $b(t) \neq 0$ for $|t| = 1$, ind $b = 0$.

(b) $t_1, \ldots, t_R$ are pairwise distinct points on the unit circle Γ.

(c) $\mathrm{Re}\,\alpha_r > -1/2$ (this ensures $|1 - t/t_r|^{2\alpha_r} \in L^1$).

(d) $-1/2 < \mathrm{Re}\,\beta_r < 1/2$, and $(-t)_{t_r}^{\beta_r}$ is defined as $\exp\{\delta_r i \times$ $\times \arg(-t/t_r)\}$, $|\arg(-t/t_r)| < \pi$ (in 2.28 this function was denoted by $\varphi_{t_r}^{\beta_r}$); so $(-t)_{t_r}^{\beta_r}$ has exactly one discontinuity (a jump) at t_r and $T[(-t)_{t_r}^{\beta_r}] \in G\mathcal{B}(l^2)$.

Note that (1) may also be written as

$$a(e^{i\vartheta}) = \prod_{r=1}^{R} (2-2\cos(\vartheta-\vartheta_r))^{\alpha_r} (-e^{i(\vartheta-\vartheta_r)})^{\beta_r}\, b(e^{i\vartheta}),$$

where $t = e^{i\vartheta}$, $t_r = e^{i\vartheta_r}$. Furthermore, due to the factorizations

$$\left|1 - \frac{t}{t_r}\right|^{2\alpha_r} = \left(1 - \frac{t}{t_r}\right)^{\alpha_r}\left(1 - \frac{t_r}{t}\right)^{\alpha_r},$$

$$(-t)_{t_r}^{\beta_r} = \left(1 - \frac{t}{t_r}\right)^{\beta_r}\left(1 - \frac{t_r}{t}\right)^{-\beta_r},$$

the function (1) allows a representation as

$$a(t) = \prod_{r=1}^{R} (1 - \frac{t}{t_r})^{\gamma_r} (1 - \frac{t_r}{t})^{\delta_r} b(t) \quad (|t| = 1), \qquad (2)$$

where $\gamma_r = \alpha_r + \beta_r$ and $\delta_r = \alpha_r - \beta_r$, and

(e) $(1 - t/t_r)^{\gamma_r}$ denotes the limit on the unit circle of that branch of the function which is analytic in $|t| < 1$ and takes the value 1 at $t = 0$; the term $(1 - t_r/t)^{\delta_r}$ is defined similarly.

7.2. The conjecture. Fisher and Hartwig conjectured that the Toeplitz determinants generated by 7.1(1),(2) have the following asymptotic behavior:

$$D_n(a) \sim G(b)^{n+1} n^q \,\tilde{E}(b; t_1, \ldots, t_R; \gamma_1, \ldots, \gamma_R; \delta_1, \ldots, \delta_R),$$

where $G(b) = \exp(\log b)_0$ (see 6.3), $\tilde{E}(\ldots)$ is some constant and

$$q = \gamma_1\delta_1 + \ldots + \gamma_R\delta_R = (\alpha_1^2 - \beta_1^2) + \ldots + (\alpha_R^2 - \beta_R^2).$$

To be precise, they also required that α_r and β_r are non-integral for all r.

This conjecture was proved for some values of α_r and β_r but its confirmation for the general case is an open problem to date.

SPECIAL CASES

7.3. Piecewise continuous generating functions. If we restrict ourselves to the case $\gamma_r = -\delta_r$ in 7.1(2) (or, equivalently, to $\alpha_r = 0$ in 7.1(1)) then we obtain just the class PC, which was considered in 6.44 and 6.50. Thus in this case the conjecture 7.2 turns out to be true and the constant $\tilde{E}(\ldots)$ is identified.

7.4. Zeros and/or poles of modulus type. Put in 7.1(2) $\gamma_r = \delta_r$ (that is, $\beta_r = 0$ in 7.1(1)). For such generating functions the conjecture 7.2 was verified in Lenard [1] for the case $b(t) \equiv 1$, $R = 2$, $\alpha_r > -1/2$, $t_1 = 1$, $t_2 = -1$ and then by Widom in full generality. Moreover, Basor succeeded in the case where only $\operatorname{Re}\gamma_r = \operatorname{Re}\delta_r$ is required (i.e. β_r in 7.1.(1) is purely imaginary). Note that a function φ of the form 7.1(d) with $\operatorname{Re}\beta_r = 0$ can be characterized by $|\varphi(t_r - 0)| = |\varphi(t_r + 0)|$. We formulate the precise result:

Let $a(t)$ be given as in 7.1, suppose $\operatorname{Re}\gamma_r = \operatorname{Re}\delta_r$ for all r and $b \in C^{1+\varepsilon}$ with some $\varepsilon > 0$. Then 7.2 holds with

$$\tilde{E}(b; t_1, \ldots, t_R; \gamma_1, \ldots, \gamma_R; \delta_1, \ldots, \delta_R)$$

$$= E(b) \prod_{r \neq s} (1 - \frac{t_s}{t_r})^{-\gamma_r\delta_s} \prod_{r=1}^{R} b_-(t_r)^{-\gamma_r} \, b_+(t_r)^{-\delta_r} \quad \times$$

$$\times \prod_{r=1}^{R} G(1+\gamma_r)G(1+\delta_r)/G(1+\gamma_r+\delta_r),$$

where $b_+ = \exp(P \log b)$, $b_- = \exp(Q \log b)$ and G is the Barnes function 6.43.

For a proof we refer to Widom [5] and Basor [1].

7.5. Zeros and/or poles of analytic type. Now suppose $\gamma_r = 0$ for all r (the case $\delta_r = 0$ is analogous). Furthermore, assume $\operatorname{Re} \delta_r > -1$ and $\operatorname{Re} \delta_r \neq 0$. For this case in Silbermann [4] the conjecture 7.2 was confirmed and precisized as follows:

If $b \in W^{0,1}$ (see 5.23(b)), where

$$1 > \delta := \max\{1, |\operatorname{Re} \delta_1|, \ldots, |\operatorname{Re} \delta_R|\},$$

$b(t) \neq 0$ for $|t| = 1$, and $\operatorname{ind} b = 0$, then

$$D_n(a) = G(b)^{n+1} \tilde{E}(b; t_1, \ldots, t_R; 0, \ldots, 0; \delta_1, \ldots, \delta_R) \times$$
$$\times \{1 + o(1/n^{1-\delta-\varepsilon})\},$$

where $\varepsilon > 0$ is an arbitrary number and

$$\tilde{E}(b; t_1, \ldots, t_R; 0, \ldots, 0; \delta_1, \ldots, \delta_R) = E(b) \prod_{r=1}^{R} b_+(t_r)^{\delta_r},$$
$$b_+ = \exp(P \log b).$$

Moreover, if $b \in W$, $b(t) \neq 0$ for $|t| = 1$, $\operatorname{ind} b = 0$, and $(b^{-1})_n = 0$ for $n \geqq n_0 \geqq 0$ then

$$D_n(a) = G(b)^{n+1} E(b) \prod_{r=1}^{R} b_+(t_r)^{\delta_r} \quad \text{for } n \geqq n_0 \geqq 0.$$

7.6. Zeros of integral orders. In general, the conjecture 7.2 fails to be true if γ_r and δ_r are non-negative integers (example: $D_n[(1+t)(1-1/t)] = 1/2 + (-1)^{n+1}/2$, while 7.2 gave $E + o(1)$ with some constant E). This is also seen from Theorem 6.29.
The rather exotic behavior of $D_n(a)$ for zeros of integral orders is described by the following result, which was established in Böttcher, Silbermann [2]:

Set $\gamma = \sum_{r=1}^{R} \gamma_r$, $\delta = \sum_{r=1}^{R} \delta_r$, $\mu_r = \gamma_r + \delta_r$, $\mu = \max_{1 \leqq r \leqq R} \{\mu_r\}$, and let $b \in W \cap l_p^{u,v}$ (cf. 4.13) where $u, v \geqq \mu + 2\varepsilon > \mu$ and $p > 1$ is an arbitrary number. Suppose $b(t) \neq 0$ for $|t| = 1$ and $\operatorname{ind} b = 0$. Without loss of generality assume $\gamma \leqq \delta$ (otherwise consider $D_n(\tilde{a}) = D_n(a)$). Define

$$\mathfrak{M} = \{m = (m_1, \ldots, m_R) \in \mathbb{Z}^R : 0 \leqq m_r \leqq \mu_r,\ m_1 + \ldots + m_R = \gamma\},$$

$$q = \max \{ \sum_{r=1}^{R} (m_r\mu_r - m_r^2) : m \in \mathfrak{M}\} ,$$

$$\mathfrak{M}^* = \{m \in \mathfrak{M} : \sum_{r=1}^{R} (m_r\mu_r - m_r^2) = q\}$$

and put $b_+ = \exp(P \log b)$, $b_- = \exp(Q \log b)$. Then

$$D_n(a) = G(b)^{n+1} G(b_-/b_+)^{\gamma} E(b)\, n^q \{1 + o(1/n^{\nu-\mu-\varepsilon})\} \times$$

$$\times \{ \sum_{m \in \mathfrak{M}^*} A_m \left(\frac{t_1^{m_1} \cdots t_R^{m_R}}{t_1^{\gamma_1} \cdots t_R^{\gamma_R}}\right)^{n+1} + O(1/n) +$$

$$+ o(1/n^{u-\mu-\varepsilon+1/p}) + O(1/n^{\nu-\mu-\varepsilon+1/p+q})\},$$

where $G(b)$, $G(b_-/b_+)$, $E(b)$ are defined as usual,

$$A_m = A_{m_1,\ldots,m_R} = \prod_{r \neq s} (1 - \frac{t_r}{t_s})^{(m_r-\mu_r)m_s} \times$$

$$\times \prod_{r=1}^{R} \frac{G(\mu_r - m_r + 1) G(m_r+1)}{G(\mu_r+1)} \prod_{r=1}^{R} b_-(t_r)^{-m_r} b_+(t_r)^{m_r-\mu_r}$$

and $G(m) = (m-2)! \ldots 2!1!0!$ $(m \geqq 2)$, $G(1) = 1$, is Barnes function 6.43.

To get an idea of the proof note that for non-negative integers γ_r and δ_r

$$a(t) = \prod_{r=1}^{R} (1 - \frac{t}{t_r})^{\gamma_r} (1 - \frac{t_r}{t})^{\delta_r} b(t)$$

$$= t^{\gamma} \prod_{r=1}^{R} (-1/t_r)^{\gamma_r} \prod_{r=1}^{R} (1 - \frac{t_r}{t})^{\gamma_r+\delta_r} b(t),$$

which allows us to apply the results of 7.5 and the Theorems 6.24 and 6.28 (see the authors' paper quoted above for details).

<u>7.7. Singularities of small order and separation.</u> In [5], Widom also proved 7.2 with $R = 1$, α,β real, $|\alpha|$, $|\beta| < 1/2$, and $b \in C^{1+|2\beta|+\varepsilon}$ $(\varepsilon > 0)$, but did not determine the constant. That is,

$$D_n[|1 - \frac{t}{t_o}|^{2\alpha} (-\frac{t}{t_o})^{\beta} b(t)] \sim \tilde{E}(b, t_o; \alpha, \beta) G(b)^{n+1} n^{\alpha^2-\beta^2} \qquad (3)$$

with some constant $\tilde{E}(\ldots) \neq 0$.

On the other hand, in Basor, Helton [1] one can find the following fact:

Let $\mathcal{E}_{50}$ denote the class of measurable functions f on Γ which factor as $f = f_- f_+$ with $f_+ \in GH^{50}$ and $f_- \in G\overline{H^{50}}$. The singular points of f are

$$\mathrm{clos}\{t \in \Gamma : \text{one of } f_{\pm}^{\pm 1} \text{ do not belong to } C \text{ near } t\}.$$

If φ and ψ are in $\mathcal{E}_{50}$ and have no common singular points, then the multiplicative commutators

$$\{T(\varphi_-), T(\psi_+)\} := T(\varphi_-)T(\psi_+)T^{-1}(\varphi_-)T^{-1}(\psi_+),$$

$$\{T(\psi_-), T(\varphi_+)\} := T(\psi_-)T(\varphi_+)T^{-1}(\psi_-)T^{-1}(\varphi_+)$$

differ from the identity only by an operator of $\mathfrak{C}_1(H^2)$ and

$$\lim_{n\to\infty} \frac{D_n(\varphi\psi)}{D_n(\varphi)D_n(\psi)} = \det\{T(\varphi_-),T(\psi_+)\} \det \{T(\psi_-),T(\varphi_+)\} \neq 0. \quad (4)$$

There is a passage in Basor, Helton [1] reading as follows: "In particular (4) implies that the high order asymptotics of $D_n(a)$ for $a(t)$ of the form 7.1(2) with $|\alpha_r|$ and $|\beta_r| < 1/400$ can be computed provided one can compute expansions for each $D_n[(1-t/t_r)^{\alpha_r-\beta_r} \times$ $\times (1-t_r/t)^{\alpha_r-\beta_r}]$ individually." Now, the latter expansions for $|\alpha_r|, |\beta_r| < 1/400$ (even for $|\alpha_r|$, $|\beta_r| < 1/2$) are given by (3), so that (4) combined with (3) would confirm the conjecture 7.2 for very small real α_r's and β_r's, but, of course, without identifying the constant.

SEPARATION OF DISJOINT SINGULARITIES

<u>7.8.</u> To prove 7.7(4) Basor and Helton used H^p-techniques (with $p \neq 2$). The approach given in the following will work with weighted H^2-spaces. As an example we shall consider the generating function

$$a(t) = \prod_{r=1}^{R_1} \left|1 - \frac{t}{t_r}\right|^{2\alpha_r} \prod_{r=R_1+1}^{R_2} (-t)_{t_r}^{\beta_r}\, b(t), \quad (5)$$

where b is smooth, $b(t) \neq 0$ for $|t| = 1$, ind $b = 0$, $t_1,\dots,t_{R_2}$ are pairwise distinct points on Γ, α_r is real, $-1/2 < \alpha_r < 1/2$ for all r, and 7.1(d) has to be taken into consideration.
We emphasize that $a(t)$ has at t_r <u>either</u> a zero or pole of modulus type <u>or</u> a jump.

First we shall define a pair of weighted H^2-spaces where the finite section method will be applicable to $T(a)$. Thereby b is only required to be continuous.
Then we shall prove a separation theorem for the determinants $D_n(a)$ generated by $a(t)$ of the form (5). The function b is then supposed to belong to B_1^1 (see 1.20, 2.17(b), 6.49).

<u>7.9. The spaces $L^2(\Gamma,\varrho)$.</u> Let $t_1,\dots,t_R$ be pairwise distinct points on the unit circle Γ and $\mu_1,\dots,\mu_R$ be real numbers. Set

$$\varrho(t) = \prod_{r=1}^{R} |t - t_r|^{\mu_R} \qquad (|t| = 1)$$

and denote by $L^2(\Gamma,\varrho)$ the Hilbert space with norm

$$\|\varphi\|_{L^2(\Gamma,\varrho)} = \left(\int_\Gamma |\varphi(t)|^2\varrho(t)|dt|\right)^{1/2} .$$

For a detailed study of these spaces see Gohberg, Krupnik [2]. We mention here only the following facts:

(a) If $-1 < \mu_r$ for all r then the linear span $\operatorname{lin}_{k\in\mathbb{Z}}\{\chi_k\}$ (see 2.4) forms a dense subspace in $L^2(\Gamma,\varrho)$.

(b) If $-1 < \mu_r < 1$ for all r then the projections P and Q defined on $\operatorname{lin}_{k\in\mathbb{Z}}\{\chi_k\}$ as

$$P\chi_k = \begin{cases}\chi_k, & k \geqq 0\\ 0, & k < 0\end{cases}, \qquad Q\chi_k = \begin{cases}0, & k \geqq 0\\ \chi_k, & k < 0\end{cases}$$

extend to bounded operators on $L^2(\Gamma,\varrho)$.

(c) If $-1 < \mu_r < 1$ for all r then the projections $\mathcal{P}_n$ $(n=0,1,2,\dots)$ onto $\operatorname{lin}\{\chi_{-n},\chi_{-n+1},\dots,\chi_n\}$ parallel to $\operatorname{clos}\operatorname{lin}_{|k|>n}\{\chi_k\}$ are bounded on $L^2(\Gamma,\varrho)$ and converge strongly to I. This is seen from the identity

$$\mathcal{P}_n\varphi = \chi_{-n}P(\chi_n\varphi) - \chi_{n+1}P(\chi_{-n-1}\varphi),\ \varphi \in L^2(\Gamma,\varrho).$$

Finally, let $\mathcal{Q}_n = I - \mathcal{P}_n$, $P_n = \mathcal{P}_nP$, $Q_n = \mathcal{Q}_nP$.

If $-1 < \mu_r < 1$ for all r then define

$$L^2_+(\Gamma,\varrho) = PL^2(\Gamma,\varrho),\ L^2_-(\Gamma,\varrho) = QL^2(\Gamma,\varrho)$$

and for $f \in L^\infty$ the operators $M(f)$ and $T_\pm(f)$ as

$$M(f) : L^2(\Gamma,\varrho) \to L^2(\Gamma,\varrho),\quad \varphi \mapsto f\varphi ; \tag{6}$$

$$T_+(f) : L^2_+(\Gamma,\varrho) \to L^2_+(\Gamma,\varrho),\quad \varphi \mapsto P(f\varphi); \tag{7}$$

$$T_-(f) : L^2_-(\Gamma,\varrho) \to L^2_-(\Gamma,\varrho),\quad \varphi \mapsto Q(f\varphi). \tag{8}$$

Note that $M(f)$, $T_+(f)$ and $T_-(f)$ are bounded if only $f \in L^\infty$. Clearly, $L^2_+(\Gamma,\varrho) = H^2$ and $L^2_-(\Gamma,\varrho) = H^2_-$ if $\varrho(t) \equiv 1$.

<u>7.10. The weight is specified.</u> Let $a(t)$ be given as in 7.8. Throughout the following the letter ϱ will be reserved to denote the weight

$$\varrho(t) = \prod_{r=1}^{R_1} |t - t_r|^{2\alpha_r} \quad (|t| = 1).$$

Since $-1/2 < \alpha_r < 1/2$ was supposed, we see that 7.9(a),(b),(c)

apply to the spaces $L^2(\Gamma,\varrho)$ and $L^2(\Gamma,\varrho^{-1})$.
Next, define on $|t| = 1$

$$\sigma_1(t) = \prod_{r=1}^{R_1}(1 - \frac{t_r}{t})^{\alpha_r}, \quad \sigma_2(t) = \prod_{r=1}^{R_1}(1 - \frac{t}{t_r})^{\alpha_r}, \tag{9}$$

$$\psi_1(t) = \prod_{r=1}^{R_1}(-t)_{t_r}^{\alpha_r}, \quad \psi_2(t) = \prod_{r=1}^{R_1}(-t)_{t_r}^{-\alpha_r}. \tag{10}$$

Then

$$\sigma_2 = \sigma_1\psi_1, \quad \sigma_1 = \sigma_2\psi_2, \quad \varrho = \sigma_1^2\psi_1 = \sigma_2^2\psi_2.$$

Finally, let the operators $M(\sigma_i^{\pm 1})$, $T_+(\sigma_i^{\pm 1})$, $T_-(\sigma_i^{\pm 1})$ be defined formally as in 7.9(6),(7),(8).

<u>7.11. Lemma.</u> Put either $X = L^2(\Gamma,\varrho)$ and $Y = L^2$ or $X = L^2$ and $Y = L^2(\Gamma,\varrho^{-1})$. Then for $i = 1,2$

$$M(\sigma_i) \in \mathcal{B}(X,Y), \quad M(\sigma_i^{-1}) \in \mathcal{B}(Y,X), \tag{11}$$

$$T_{\pm}(\sigma_i) \in \mathcal{B}(X_{\pm},Y_{\pm}), \quad T_{\pm}(\sigma_i^{-1}) \in \mathcal{B}(Y_{\pm},X_{\pm}). \tag{12}$$

Moreover,

$$M(\sigma_i)M(\sigma_i^{-1}) = I_Y, \quad M(\sigma_i^{-1})M(\sigma_i) = I_X, \tag{13}$$

$$T_{\pm}(\sigma_i)T_{\pm}(\sigma_i^{-1}) = I_{Y_{\pm}}, \quad T_{\pm}(\sigma_i^{-1})T_{\pm}(\sigma_i) = I_{X_{\pm}}. \tag{14}$$

<u>Proof.</u> For definiteness, take $X = L^2(\Gamma,\varrho)$ and $Y = L^2$. Since $|\psi_i(t)| = 1$, we have

$$\int_\Gamma |\sigma_i\varphi|^2|dt| = \int_\Gamma |\varphi|^2|\sigma_i^2\psi_i||dt| = \int_\Gamma |\varphi|^2\varrho|dt|,$$

hence $M(\sigma_i) \in \mathcal{B}(L^2(\Gamma,\varrho),L^2)$. Equally, $M(\sigma_i^{-1}) \in \mathcal{B}(L^2,L^2(\Gamma,\varrho))$. This proves (11) and (13). (12) is due to 7.9(b) an obvious consequence of (11). Lastly, the identity

$$T_+(\sigma_i)T_+(\sigma_i^{-1})\varphi = PM(\sigma_i)PM(\sigma_i^{-1})P\varphi$$

is easily checked for $\varphi = \chi_k$, $k \in \mathbb{Z}_+$, and extends to all $\varphi \in L_+^2$ because of the density of $\operatorname{lin}_{k\in\mathbb{Z}_+}\{\chi_k\}$ in L_+^2 (cf. 7.9(a),(b)). The other assertions of (14) can be shown analogously. Δ

Henceforth $T_+(f)$ will simply be denoted by $T(f)$.

<u>7.12. Corollary.</u> Let $a(t)$ be defined by 7.8(5) and suppose $b \in C$. Then

$$T(a) : L_+^2(\Gamma,\varrho) \to L_+^2(\Gamma,\varrho^{-1})$$

is bounded and invertible.

Proof. Put

$$c(t) = b(t) \prod_{r=R_1+1}^{R_2} (-t)_{t_r}^{\beta_r} \qquad (|t| = 1). \tag{15}$$

Then the diagram

$$\begin{array}{ccccccc} L_+^2(\Gamma,\varrho) & \xrightarrow{T(\sigma_2)} & L_+^2 & \xrightarrow{T(c)} & L_+^2 & \xrightarrow{T(\sigma_1)} & L_+^2(\Gamma,\varrho^{-1}) \\ \lfloor & \multicolumn{5}{c}{\underrightarrow{\qquad\qquad\qquad T(a) \qquad\qquad\qquad}} & \uparrow \end{array}$$

is commutative, $T(\sigma_2)$ and $T(\sigma_1)$ are bounded and invertible by Lemma 7.11, while $T(c)$ is so by 2.29 and 2.8. Δ

7.13. Theorem. If $a(t)$ is the function 7.8(5), with $b \in C$, then (cf. 3.1)

$$T(a) \in \Pi_{L_+^2(\Gamma,\varrho),L_+^2(\Gamma,\varrho^{-1})}\{P_nT(a)P_n\}.$$

Proof. By virtue of Corollary 7.12 and Proposition 3.8 it suffices to show the invertibility of

$$Q_nT^{-1}(a)Q_n : \operatorname{Im} Q_n \to \operatorname{Im} Q_n$$

for all n large enough (say $n > n_o$) and

$$\sup_{n>n_o} \|(Q_nT^{-1}(a)Q_n)^{-1}Q_n\|_{\varrho \to \varrho^{-1}} < \infty .$$

But

$$T^{-1}(a) = T^{-1}(\sigma_2)T^{-1}(c)T^{-1}(\sigma_1) = T(\sigma_2^{-1})T^{-1}(c)T(\sigma_1^{-1}) =$$

$$= T(\sigma_1^{-1})T(\psi_2)T^{-1}(c)T(\psi_1)T(\sigma_2^{-1}),$$

hence

$$Q_nT^{-1}(a)Q_n = Q_nT(\sigma_1^{-1})Q_nT(\psi_2)T^{-1}(c)T(\psi_1)Q_nT(\sigma_2^{-1})Q_n.$$

From Lemma 7.11 and 7.9(c) we know that

$$\sup_{n \geqq 0}\|(Q_nT(\sigma_1^{-1})Q_n)^{-1}Q_n\|_{\varrho \to 1} \leqq \|T(\sigma_1)\|_{\varrho \to 1} \sup_{n \geqq 0}\|Q_n\|_{\varrho \to \varrho} < \infty ,$$

$$\sup_{n \geqq 0}\|(Q_nT(\sigma_2^{-1})Q_n)^{-1}Q_n\|_{1 \to \varrho^{-1}} \leqq \sup_{n \geqq 0}\|Q_n\|_{\varrho^{-1} \to \varrho^{-1}}\|T(\sigma_2)\|_{1 \to \varrho^{-1}} < \infty,$$

and so we are left with the operator

$$Q_nT(\psi_1)T^{-1}(c)T(\psi_1)Q_n + P_n : H^2 \to H^2. \tag{16}$$

Now $T(\psi_2)T(\psi_1) = I - PM(\psi_2)QM(\psi_1)P$. Since ψ_1 is unimodular and $T(\psi_1) \in G\mathcal{B}(H^2)$, there is a function $h \in H^\infty$ such that $\|\psi_1 - h\|_{L^\infty} < 1$ (see Theorem 2.15). Consequently, $\|PM(\psi_2)QM(\psi_1)P\| < 1$, which gives the invertibility of $Q_nT(\psi_2)T(\psi_1)Q_n + P_n$ on H^2 and the uniform boundedness of the norms of the inverses. This and Proposition 3.8 yield $T^{-1}(\psi_1)T^{-1}(\psi_2) \in \Pi_2\{P_n\}$. Furthermore, $T(c) \in \Pi_2\{P_n\}$ by Corollary 2 in 3.18. Now Theorem 3.16 can be applied to conclude that

$$T^{-1}(\psi_1)T(c)T^{-1}(\psi_2) \in \Pi_2\{P_n\}; \tag{17}$$

Indeed, the operator on the left of (17) is invertible, it is not too difficult to check that

$$\text{s-}\lim_{n\to\infty} W_n T^{-1}(\psi_1)T(c)T^{-1}(\psi_2)W_n = T(\tilde{c}),$$

and the invertibility of the corresponding coset in $\mathfrak{A}^0$ can be shown with the help of the local principle in the same fashion as in 3.19 (with $p = 2$) - we emphasize once more that ψ_1 and c as well as ψ_2 and c have no common singularities.

Finally, again with the aid of Proposition 3.8 we can deduce from (17) that the operators (16) are invertible for all sufficiently large n and that the norms of their inverses are uniformly bounded. This completes the proof. Δ

<u>Remark.</u> This theorem can be generalized to the case where $a(t)$ additionally involves factors of the form

$$\prod_{r=R_2+1}^{R_3}\left(1-\frac{t}{t_r}\right)^{\gamma_r}, \quad \prod_{r=R_3+1}^{R_4}\left(1-\frac{t_r}{t}\right)^{\delta_r} \quad (|t| = 2)$$

with $|\mathrm{Re}\,\gamma_r|$, $|\mathrm{Re}\,\delta_r| < 1$, and the points $t_1,\ldots,t_{R_4}$ being pairwise distinct. Then the weight ϱ must be modified in such a way that again

$$T(a) : L^2_+(\Gamma,\varrho) \to L^2_+(\Gamma,\varrho^{-1})$$

be bounded and invertible.

<u>7.14. Theorem (Verbickiĭ).</u> Let c be given by 7.12(1) with $b \in C$. Then

$$T(c) \in \Pi_{L^2_+(\Gamma,\varrho)}\{P_n\} \quad \text{and} \quad T(c) \in \Pi_{L^2_+(\Gamma,\varrho^{-1})}\{P_n\}.$$

<u>Proof.</u> Verbickiĭ [3], Theorem 4. Δ

Now we turn over to the asymptotic behavior of the Toeplitz determinants generated by the function 7.8(5). The function b is henceforth required to be in B^1_1.

<u>7.15. Theorem.</u> Write the function 7.8(5) in the form $a = \sigma_1\sigma_2 c$ with $\sigma_1\sigma_2$ and c defined by 7.10(9) and 7.12(15), respectively. Let ψ_1 and ψ_2 denote the functions 7.10(10).

Then

$$F(\psi_1,c)-I := T^{-1}(c)T^{-1}(\psi_1)T(\psi_1c)-I \in \mathfrak{C}_1(H^2),$$

$$F(\tilde{\psi}_2,\tilde{c})-I := T^{-1}(\tilde{c})T^{-1}(\tilde{\psi}_2)T(\tilde{\psi}_2c)-I \in \mathfrak{C}_1(H^2)$$

and

$$\lim_{n\to\infty} \frac{D_n(a)}{D_n(\sigma_1\sigma_2)D_n(c)} = \det F(\psi_1,c) \cdot \det F(\tilde{\psi}_2,\tilde{c}) \neq 0. \quad (18)$$

<u>Proof.</u> The operators $M(c)$ and $M(\sigma_1\sigma_2)$ are thought of as acting on the following spaces:

$$M(c) : L^2(\Gamma,\varrho) \to L^2(\Gamma,\varrho),\ M(\sigma_1\sigma_2) : L^2(\Gamma,\varrho) \to L^2(\Gamma,\varrho^{-1}).$$

Furthermore, writing in the following $H(f)$ and $H(\tilde{f})$ we actually mean PfQ and QfP, respectively (contrary to 2.3, where $H(.)$ still involved the operator J). Then it is clear how the operators $T(c) = PM(c)P|\operatorname{Im} P$, $H(\tilde{c}) = QM(c)P$, $T(\sigma_1\sigma_2) = PM(\sigma_1\sigma_2)P$ etc. are to be understood.

Now abbreviate $P_nT(f)P_n|\operatorname{Im} P_n$ to $T_n(f)$ and notice the analogue of Proposition 3.6:

$$\begin{aligned} T_n(a) &= T_n(\sigma_1\sigma_2c) \\ &= T_n(\sigma_1\sigma_2)T_n(c) + P_nH(\sigma_1\sigma_2)H(\tilde{c})P_n + W_nH(\tilde{\sigma}_1\tilde{\sigma}_2)H(c)W_n. \end{aligned}$$

From writing $W_nH(\tilde{\sigma}_1\tilde{\sigma}_2)H(c)W_n \in \mathfrak{B}(L^2_+(\Gamma,\varrho),\ L^2_+(\Gamma,\varrho^{-1}))$ as $P_nT(\sigma_1\sigma_2)Q_nT(c)P_n$ we see that it converges strongly to 0. Taking into account that $T(\sigma_1\sigma_2) \in \prod_{\varrho,\varrho^{-1}}\{P_n\}$ by Theorem 7.13, and $T(c) \in \prod_{\varrho,\varrho}\{P_n\}$ by Theorem 7.14, we get for sufficiently large n

$$K_n = P_n + A_n + B_n,$$

where

$$\begin{aligned} K_n &= T_n^{-1}(c)T_n^{-1}(\sigma_1\sigma_2)T_n(a), \\ A_n &= T_n^{-1}(c)T_n^{-1}(\sigma_1\sigma_2)P_nH(\sigma_1\sigma_2)H(\tilde{c})P_n, \\ B_n &= T_n^{-1}(c)T_n^{-1}(\sigma_1\sigma_2)W_nH(\tilde{\sigma}_1\tilde{\sigma}_2)H(c)W_n. \end{aligned}$$

Thus

$$K_n = (P_n + B_n)(P_n + A_n) - B_nA_n. \quad (19)$$

Now define $\psi(t) := \prod_{r=R_1+1}^{R_2} (-t)_{t_r}^{\beta_r}$ $(|t| = 1)$. Then $c = b\psi$ and

$$H(\sigma_1\sigma_2)H(\tilde{c}) = PM(\sigma_1\sigma_2)QM(c)P = PM(\sigma_1^2)PM(\psi_1)QM(c)P =$$

$$= PM(\sigma_1^2)PM(\psi_1)QM(\psi)PM(b)P + PM(\sigma_1^2)PM(\psi_1)QM(\psi)QM(b)P =$$

$$= T^2(\sigma_1)H(\psi_1)H(\tilde{\psi})T(b) + T^2(\sigma_1)H(\psi_1)T_-(\psi)H(\tilde{b}). \tag{20}$$

As in the proof of Proposition 2.26 one can show that $H(\psi_1)H(\tilde{\psi}) \in \mathfrak{C}_1(L_+^2(\Gamma,\varrho))$ and since $b \in B_1^1$, we have $H(\tilde{b}) \in \mathfrak{C}_1(H^2,H_-^2)$. But

$$H(\tilde{b})T(\sigma_2) = QM(b)PM(\sigma_2)P = QM(b\sigma_2)P =$$

$$= QM(\sigma_2)Q \cdot QM(b)P = T_-(\sigma_2)H(\tilde{b}),$$

i.e. $H(\tilde{b}) = T_-^{-1}(\sigma_2)H(\tilde{b})T(\sigma_2)$, and by Lemma 7.11, $T_-^{-1}(\sigma_2) = T_-(\sigma_2^{-1}) \in \mathfrak{B}(H_-^2, L_-^2(\Gamma,\varrho))$, $T(\sigma_2) \in \mathfrak{B}(L_+^2(\Gamma,\varrho),H^2)$, so that even $H(\tilde{b}) \in \mathfrak{C}_1(L_+^2(\Gamma,\varrho), L_-^2(\Gamma,\varrho))$. Thus by (20),

$$H(\sigma_1\sigma_2)H(\tilde{c}) \in \mathfrak{C}_1(L_+^2(\Gamma,\varrho), L_+^2(\Gamma,\varrho^{-1}))$$

and 1.3(d) implies that B_nA_n converges in the norm of $\mathfrak{C}_1(L_+^2(\Gamma,\varrho))$ to 0. It follows also from 1.3(d) that A_n in the norm of $\mathfrak{C}_1(L_+^2(\Gamma,\varrho))$ tends to

$$T^{-1}(c)T^{-1}(\sigma_1\sigma_2)H(\sigma_1\sigma_2)H(\tilde{c}).$$

Analogously can be shown that $W_n\ B_n\ W_n$ converges in the norm of $\mathfrak{C}_1(L_+^2(\Gamma,\varrho))$ to

$$T^{-1}(\tilde{c})T^{-1}(\tilde{\sigma}_1\tilde{\sigma}_2)H(\tilde{\sigma}_1\tilde{\sigma}_2)H(c).$$

Thus (19) gives

$$D_n(a)/(D_n(\sigma_1\sigma_2)D_n(c)) = \det K_n =$$

$$= \det(P_n + A_n) \cdot \det(P_n+W_nB_nW_n) + o(1)$$

$$= \det(I + T^{-1}(c)T^{-1}(\sigma_1\sigma_2)H(\sigma_1\sigma_2)H(\tilde{c})) \times$$

$$\times \det(I + T^{-1}(\tilde{c})T^{-1}(\tilde{\sigma}_1\tilde{\sigma}_2)H(\tilde{\sigma}_1\tilde{\sigma}_2)H(c)) + o(1),$$

where the occuring operator determinants are referred to $L_+^2(\Gamma,\varrho)$. Furthermore,

$$\det(I + T^{-1}(c)T^{-1}(\sigma_1\sigma_2)H(\sigma_1\sigma_2)H(\tilde{c}))$$

$$= \det(I + T^{-1}(c)T^{-1}(\sigma_1\sigma_2)(T(\sigma_1\sigma_2c) - T(\sigma_1\sigma_2)T(c)))$$

$$= \det T^{-1}(c)T^{-1}(\sigma_1\sigma_2)T(\sigma_1\sigma_2c)$$

$$= \det T^{-1}(c)T^{-1}(\psi_1)T^{-1}(\sigma_1^2)T^{-1}(\sigma_1^2)T(\psi_1c)$$

$$= \det T^{-1}(c)T^{-1}(\psi_1)T(\psi_1c) = \det F(\psi_1,c)$$

and similarly the other determinant can be identified with $\det F(\tilde{\psi}_2,\tilde{c})$. But from the identity

$$T^{-1}(f)T^{-1}(g)T(gf) = I + T^{-1}(f)T^{-1}(g)H(g)H(\tilde{f})$$

and from Proposition 2.26 we deduce, that $F(\psi_1,c) - I$ and $F(\tilde{\psi}_2,\tilde{c})-I$ are also in $\mathfrak{C}_1(H^2)$. Finally, 1.3(d) combined with 1.8(a) gives

$$\det T^{-1}(f)T^{-1}(g)T(gf) = \lim_{n\to\infty} P_n T^{-1}(f)T^{-1}(g)T(fg)P_n,$$

which implies that $\det F(\psi_1,c)$ and $\det F(\tilde{\psi}_2,\tilde{c})$ take the same value when regarded on H^2 instead of on $L^2_+(\Gamma,\varrho)$. This proves (18). Δ

Remark 1. With the methods of 6.42, 6.45 - 6.50 the constants $\det F(\psi_1,c)$ and $\det F(\tilde{\psi}_2,\tilde{c})$ can be determined. What results is

$$\det F(\psi_1,c) \cdot \det F(\tilde{\psi}_2,\tilde{c})$$
$$= \prod_{s=1}^{R_1} b_-(t_s)^{-\alpha_s} b_+(t_s)^{-\alpha_s} \prod_{s=1}^{R_1} \prod_{r=R_1+1}^{R_2} (1-\frac{t_s}{t_r})^{-\alpha_s\beta_r}(1-\frac{t_r}{t_s})^{\alpha_s\beta_r}. \quad (21)$$

Putting this together with the results of 7.4 and 6.50 we get for the generating function 7.8(5) (with $b \in B_1^1$) the following:

$$D_n(a) \sim G(b)^{n+1}E(b)n^q \prod_{r=1}^{R_2} b_-(t_r)^{-(\alpha_r+\beta_r)} b_+(t_r)^{-(\alpha_r-\beta_r)} \times$$
$$\times \prod_{r=1}^{R_2} \frac{G(1+\alpha_r+\beta_r)G(1+\alpha_r-\beta_r)}{G(1+2\alpha_r)} \prod_{1\leq r\neq s\leq R_2} (1-\frac{t_s}{t_r})^{-(\alpha_r+\beta_r)(\alpha_s-\beta_s)},$$

where $G(b)$, $E(b)$ are defined as usual, $b_- = \exp(Q \log b)$, $b_+ = \exp(P \log b)$, $q = \alpha_1^2 +\ldots+ \alpha_{R_1}^2 - (\beta_{R_1+1}^2 +\ldots+ \beta_{R_2}^2)$, G is Barnes function 6.43 and α_s and β_r have to be set 0 for $s = R_1+1,\ldots R_2$ and $r = 1,\ldots,R_1$, respectively.

Remark 2. Note that Theorem 7.15 remains true for continuous b belonging to B_1^1 only in a neighborhood of the points t_r $(r = 1,\ldots,R_2)$. Apart from minor modifications this can be proved as above for the case where $b \in B_1^1$ (globally). Then asymptotic formulas like those given in the corollaries in 6.44 can be established, of course, without identifying the constants.

Remark 3. The arguments given here also apply to the generating functions mentioned in the remark in 7.13. Then, however, we must require that either $|\mathrm{Re}\,\gamma_r|$, $|\mathrm{Re}\,\delta_r| < 1/2$ or $\alpha_r = 0$ for all r. But it seems that this effect is not in the nature of the matter, but rests upon the techniques used here, which are not yet flexible enough.

On the other hand, it becomes clear that further progress in this direction can be made, once the finite section method in spaces with weight has been studied in detail.

NOTES AND COMMENTS

<u>7.1. - 7.7.</u> References are given in the text.

<u>7.8. - 7.15.</u> These results are due to the authors and are published here for the first time. In comparization with the separation theorem quoted in 7.7 note that Theorem 7.15 is not restricted to singularities of very small order (i.e. in applying it to the class of Fisher and Hartwig a condition of type $|\alpha_r|, |\beta_r| < 1/400$ is not needed). Moreover, requiring only $b \in B_1^1$ the minimal smoothness of b seems to be attained. Finally, the expression (21) for the constant is simpler than that given for the right hand side of (4) in Basor, Helton [1].

Lastly, note that the method of the proof of Theorem 7.13 yields an asymptotic inverse of $T_n(a)$ (cf. in this connection Blecher [1]).

REFERENCES

Achieser, N.I. (Ахиезер, Н.И.)

[1] Континуальный аналог некоторых теорем о теплицевых матрицах. Укр. мат. ж., т. I6 (I964), № 4, 445-462.

Adamian, V.M., D.Z. Arov, M.G. Krein (Адамян, В.М., Д.З. Аров, М.Г. Крейн)

[1] Аналитические свойства пар Шмидта ганкелева оператора и обобщенная задача Шура-Такаги. Мат. сборник 86 (I97I), № I, 34-75.

Ambarcumian, G.V. (Амбарцумян, Г.В.)

[1] О методе редукции для одного класса теплицевых матриц. Мат. иссл. 8 (I973), № 2, I6I-I69.

Axler, S., S.-Y.A. Chang, and D. Sarason

[1] Products of Toeplitz operators. Integral Equations and Operator Theory 1 (1978), No.3, 285-309.

Barnes, E.W.

[1] The theory of the G-function. Quart. J. Pure Appl. Math. 31 (1900), 264-313.

Basor, E.

[1] Asymptotic formulas for Toeplitz determinants. Trans. Amer. Math. Soc. 239 (1978), 33-65.

[2] A localization theorem for Toeplitz determinants. Indiana Univ. Math. J. 28 (1979), 975-983.

[3] Asymptotic formulas for Toeplitz and Wiener-Hopf operators. Integral Equations and Operator Theory 5 (1982), No. 5, 659-672.

Basor, E. and J.W. Helton

[1] A new proof of the Szegö limit theorem and new results for Toeplitz operators with discontinuous symbol. J. Operator Theory 3 (1980), 23-39.

Baxter, G.

[1] A convergence equivalence related to polynomials orthogonal on the unit circle. Trans. Amer. Math. Soc. 99 (1961), 471-487.

[2] Polynomials defined by a difference system. J. Math. Anal. Appl. No.2 (1961), 223-263.

[3] A norm inequality for a finite-section Wiener-Hopf equation. Illinois J. Math. 7 (1963), 97-103.

Baxter, G. and I.I. Hirschman, Jr.

[1] An explicit inversion formula for finite-section Wiener-Hopf operators. Bull. Amer. Math. Soc. 70 (1964), 820-823.

Bergh, J. and J. Löfström

[1] Interpolation spaces. An introduction. Springer-Verlag, 1976

Blecher, P.M. (Блехер, П.М.)

[1] Обращение теплицевых матриц. Труды Моск. Мат. Общества, т. 40 (I979), 207-240.

[2] О гипотезе Фишера-Хартвига в теории теплицевых матриц. Функц. анализ и его прилож. I6 (I982), вып.2, I-6.

Böttcher, A.

[1] Toeplitz determinants with piecewise continuous generating function. Z. f. Analysis u. Anw. 2 (1982), 23-39.

[2] Das Reduktionsverfahren für nichtelliptische Wiener-Hopf'sche Integraloperatoren in einer Klasse von topologischen Vektorräumen. Wiss. Z. d. TH Karl-Marx-Stadt 25 (1983), Heft 3

[3] The finite section method for two-dimensional Wiener-Hopf integral operators in L^p with piecewise continuous symbols. Math. Nachr. (to appear)

[4] Двумерные свертки в углах с ядрами, имеющими носитель в полуплоскости. Мат. заметки (в печати)

[5] О нётеровости и редукции двумерных операторов Винера-Хопфа с кусочно-непрерывными символами. ДАН СССР (в печати)

Böttcher, A. and B. Silbermann

[1] Notes on the asymptotic behavior of block Toeplitz matrices and determinants. Math. Nachr. 98 (1980), 183-210.

[2] The asymptotic behavior of Toeplitz determinants for generating functions with zeros of integral orders. Math. Nachr. 102 (1981), 79-105.

[3] Ueber das Reduktionsverfahren für diskrete Wiener-Hopf-Gleichungen mit unstetigem Symbol. Z. f. Analysis u. Anw. 2 (1982), 1-5.

[4] The finite section method for Toeplitz operators on the quarter-plane with piecewise continuous symbols. Math. Nachr. (to appear)

Brown, A. and P.R. Halmos

[1] Algebraic properties of Toeplitz operators. J. reine u. angew. Math. 231 (1963), 89-102.

Calderon, A., F. Spitzer, and H. Widom

[1] Inversion of Toeplitz matrices. Illinois J. Math. 3 (1959), 542-559.

Clancey, K.

[1] Exact sequences of algebras generated by singular integral operators. Integral Equations and Operator Theory 4 (1981), 185-205.

Clancey, K. and I. Gohberg

[1] Localization of singular integral operators. Math. Z. 169 (1979), 105-117.

[2] Factorization of matrix functions and singular integral operators. Birkhäuser Verlag, 1981.

Coburn, L.A.

[1] Weyl's theorem for nonnormal operators. Michigan Math. J. 13 (1966), No.3, 285-288.

[2] The C*-algebra generated by an isometry, I. Bull. Amer. Math. Soc. 73 (1967), 722-726.

Day, K.M.

[1] Toeplitz matrices generated by the Laurent series expansion of an arbitrary rational function. Trans. Amer. Math. Soc. 206 (1975), 224-245.

Devinatz, A.

[1] Toeplitz operators on H^2 spaces. Trans. Amer. Math. Soc. 112 (1964), 304-317.

[2] The strong Szegö limit theorem. Illinois J. Math. 11 (1967) 160-175.

Douglas, R.G.

[1] Toeplitz and Wiener-Hopf operators in $H^\infty + C$. Bull. Amer. Math. Soc. 74 (1968), 895-899.

[2] Banach algebra techniques in operator theory. Academic Press, New York, 1972.

[3] Banach algebra techniques in the theory of Toeplitz operators. C.B.M.S. Lecture Notes, No.15, Amer. Math. Soc. Providence, R.I., 1973.

[4] Local Toeplitz operators. Proc. Lond. Math. Soc., 3rd ser., 36 (1978), 243-272.

Douglas, R.G. and R. Howe

[1] On the C*-algebra of Toeplitz operators on the quarter-plane. Trans. Amer. Math. Soc. 158 (1971), 203-217.

Dudučava, R. V. (Дудучава, Р.В.)

[1] Дискретные уравнения Винера-Хопфа в пространстве ℓ^p с весом. Сообщ. АН Груз. ССР 67 (1972), № 1, 17-20.

[2] Дискретные уравнения Винера-Хопфа, составленные из коэффициентов Фурье кусочно-винеровских функций. ДАН СССР 208 (1972), № 1, 19-22.

[3] О дискретных уравнениях Винера-Хопфа. Труды Тбилисского Мат. Института, т. 50 (1975), 42-59.

[4] Интегральные операторы свертки на квадранте с разрывными символами. Изв. АН СССР, сер. мат., 40 (1976), № 2, 388-412.

[5] Дискретные операторы свертки на квадранте и их индексы. Изв. АН СССР, сер. мат., 41 (1977), № 5, 1125-1137.

[6] Integral equations with fixed singularities. Teubner-Text, Leipzig, 1979.

[7] О бисингулярных интегральных операторах и операторах свертки на квадранте. ДАН СССР 221 (1975), № 2, 279-282.

Dunford, N. and J. Schwartz

[1] Linear operators, Part II, Spectral theory. Interscience, New York, 1963.

Duren, P.L.

[1] Theory of H^p spaces. Academic Press, New York and London, 1970.

Dybin, V.B. and A.V. Kozak (Дыбин, В.Б. и А.В.Козак)

[1] Операторы Теплица: нётеровость, обратимость и редукция. Ростов-на-Дону, неопубликовано, 1980.

Dym, H.

[1] Trace formulas for a class of Toeplitz-like operators, II. J. Funct. Analysis 28 (1978), No.1, 33-57.

[2] Trace formulas for blocks of Toeplitz-like operators. J. Funct. Analysis 31 (1979), No.1, 69-100.

Fisher, M.E.

[1] Statistical mechanics of dimers on a plane lattice. Phys. Review 124 (1961), No.6, 1664-1672.

Fisher, M.E. and R.E. Hartwig

[1] Toeplitz determinants: some applications, theorems and conjectures. Adv. Chem. Phys. 15 (1968), 333-353.

[2] Asymptotic behavior of Toeplitz matrices and determinants. Arch. Rat. Mech. Anal. 32 (1969), 190-225.

Fisher, M.E. and J. Stephenson

[1] Statistical mechanics of dimers on a plane lattice, II, Dimer correlations and monomers. Phys. Review 132 (1963), No.4, 1411-1431.

Friedlander, B., M. Morf, T. Kailath and L. Ljung

[1] New inversion formulas for matrices classified in terms of their distance from Toeplitz matrices. Lin. Alg. and its Appl. 27 (1979), 31-60.

Gantmacher, F.R.

[1] Matrizenrechnung, I. Deutscher Verlag d. Wiss., Berlin, 1958.

Gohberg, I.C. (Гохберг, И.Ц.)

[1] Об одном применении теории нормированных колец к сингулярным интегральным уравнениям. УМН 7 (1952), вып.2, 149-156.

[2] О теплицевых матрицах, составленных из коэффициентов Фурье кусочно-непрерывных функций. Функц. анализ и его прилож. 1 (1967), 91-92.

Gohberg, I.C. and I.A. Feldman (Гохберг, И.Ц. и И.А. Фельдман)

[1] Уравнения в свертках и проекционные методы их решения. Наука, Москва, 1971. (English transl.: Convolution equations and projection methods for their solution. Transl. Math. Monographs, vol. 41, Amer. Math. Soc., Providence, R.I., 1974)

Gohberg, I.C. and G. Heinig (Гохберг, И.Ц. и Г. Хайниг)

[1] Обращение конечных теплицевых матриц, составленных из элементов некоммутативной алгебры. Revue Romaine de Mathematiques pures et appliquees 19 (1974), No.5, 623-663.

Gohberg, I.C. and M.G. Krein (Гохберг, И.Ц. и М.Г. Крейн)

[1] Системы интегральных уравнений на полупрямой с ядрами, зависящими от разности аргументов. УМН 13 (1958), вып.2, 3-72.

[2] Введение в теорию линейных несамосопряженных операторов в гильбертовом пространстве. Наука, Москва, 1965. (English transl.: Introduction to the theory of linear nonselfadjoint operators in Hilbert space. Transl. Math. Monographs, vol 18, Amer. Math. Soc., Providence, R.I., 1969.)

Gohberg, I.C. and N. Ya. Krupnik (Гохберг, И.Ц. и Н.Я. Крупник)

[1] Об алгебре, порожденной теплицевыми матрицами. Функц. анализ и его прилож. 3 (1969), вып.2, 46-56.

[2] Введение в теорию одномерных сингулярных интегральных операторов. Штиинца, Кишинев, 1973. (German trans.: Einführung in die Theorie der eindimensionalen singulären Integraloperatoren. Birkhäuser Verlag, 1979.)

Gohberg, I.C. and A.A. Semencul (Гохберг, И.Ц. и А.А. Семенцул)

[1] Об обращении конечных теплицевых матриц и их континуальных аналогов. Мат. иссл. 7 (1972), № 2, 201-224.

Golinskiĭ, B.L. and I.A. Ibragimov (Голинский, Б.Л. и И.А. Ибрагимов)

[1] О предельной теореме Г.Сегё. Изв. АН СССР, сер. мат., 35 (1971), № 2, 408-427.

Gorodeckiĭ, M.B. (Городецкий, М.Б.)

[1] О нётеровости и редукции многомерных дискретных сверток. Изв. вузов, Математика, № 4(227) (1981), 12-15.

[2] Определители Теплица, порожденные рациональными функциями. Деп. в ВИНИТИ, № 5451-81 Деп., РЖ Мат. 1982, 3А444 Деп.

Grenander, U. and G. Szegö

[1] Toeplitz forms and their applications. Univ. of California Press, 1958.

Guillemin, V.

[1] Some classical theorems in spectral theory revisited. In: Seminar on Singularities of Solutions of Linear Partial Differential Equations, Princeton Univ. Press, 1979, 219-259.

Halmos, P.R.

[1] A Hilbert space problem book. D. Van Nostrand Company, Inc., Princeton, N.J., 1967.

Hartman, P.

[1] On completely continuous Hankel matrices. Proc. Amer. Math. Soc. 9 (1958), No.6, 862-866.

Hartman, P. and A. Wintner

[1] The spectra of Toeplitz's matrices. Amer. J. Math. 76 (1954), 867-882.

Hartwig, R.E.

[1] Monomer pair correlations. J. Math. Phys. 7 (1966), No.2, 286-299.

Heinig, G.

[1] Endliche Toeplitzmatrizen und zweidimensionale Wiener-Hopf-Operatoren mit homogenem Symbol, I, II. Math. Nachr. 82 (1978), 29-52, 53-68.

[2] Struktur des Kernes und partielle Indizes bei allgemeinen linearen Operatoren. Preprint P-36/80, Akad. Wiss. DDR,ZIMM, 1980.

[3] Beiträge zur Spektraltheorie von Operatorbüscheln und zur algebraischen Theorie von Toeplitzmatrizen. Diss. B, TH Karl-Marx-Stadt, 1979.

[4] Inversion of Toeplitz matrices with singular sections. Wiss. Z. d. TH Karl-Marx-Stadt 25 (1983), Heft 4.

Heinig, G. and K. Rost

[1] Invertierung einiger Klassen von Matrizen und Operatoren, I. Endliche Toeplitzmatrizen und ihre Verallgemeinerungen. Wiss. Informationen 12, TH Karl-Marx-Stadt, 1979.

[2] Invertierung von Toeplitzmatrizen und ihrer Verallgemeinerungen,I. Die Methode der UV-Reduktion. Beiträge zur num.Math.(to appear).

Heinig, G. and B. Silbermann

[1] Factorization of matrix functions in algebras of bounded functions. Preprint P-MATH-05/83, Akad. Wiss. DDR, Inst. f. Math., 1983.

Helson, H. and G. Szegö

[1] A problem in prediction theory. Ann. Math. Pura Appl. 51 (1960), 107-138.

Helton, J.W. and R.E. Howe

[1] Integral operators: traces, index, and homology. In: Proceedings of the conference on operator theory, Lecture Notes in Math., Vol.345, Springer-Verlag, 1973, 141-209.

Hirschman, I.I.,Jr.

[1] On a formula of Kac and Achieser.J. Math. Mech. 16 (1966), 167-196.

Hirschman, I.I., Jr. and D.E. Hughes

[1] Extreme eigenvalues of Toeplitz operators. Lecture Notes in Math., Vol.618, Springer-Verlag, 1977.

Hoffman, K.

[1] Banach spaces of analytic functions. Prentice-Hall, Englewood Cliffs, N.J., 1962.

Høholdt, T. and J. Justesen

[1] Determinants of a class of Toeplitz matrices. Math. Scand. 43 (1978), 250-258.

Iochvidov, I.S. (Иохвидов, И.С.)

[1] Ганкелевы и теплицевы матрицы и формы. Наука, Москва, I974.

Junghanns, P. and B. Silbermann

[1] Lokale Theorie des Kollokationsverfahrens zur näherungsweisen Lösung singulärer Integralgleichungen. Preprint P-MATH-10/82, Akad. Wiss. DDR, Inst. f. Math., 1982.

Kac, M.

[1] Toeplitz matrices, translation kernels and a related problem in probability theory. Duke Math. J. 21 (1954), 501-510.

Kailath, T., S.-Y. Kung, and M. Morf

[1] Displacement ranks of matrices and linear equations. J. of Math. Analysis and Appl. 68 (1979), No.2, 395-407.

Kailath, T., A. Vieira, and M. Morf

[1] Inverses of Toeplitz operators, innovations and orthogonal polynomials. SIAM-Review, Jan. 1978, 106-119.

Köhler, U. and B. Silbermann

[1] Einige Ergebnisse über Φ_+-Operatoren in lokalkonvexen topologischen Vektorräumen. Math. Nachr. 56 (1973), 145-153.

[2] Ueber algebraische Eigenschaften einer Klasse von Operatormatrizen und eine Anwendung auf singuläre Integraloperatoren. Math. Nachr. 57 (1973), 245-258.

Korneĭčuk, N.P. (Корнейчук, Н.П.)

[1] Экстремальные задачи теории приближения. Наука, Москва, I976.

Kozak, A.V. (Козак, А.В.)

[1] О методе редукции для многомерных дискретных сверток. Мат. иссл. VIII (I973), вып. 3(29), I57-I60.

[2] Локальный принцип в теории проекционных методов. ДАН СССР 2I2 (I973), № 6, I287-I289.

Kozak, A.V. and I.B. Simonenko (Козак, А.В. и И.Б. Симоненко)

[1] Проекционные методы решения многомерных дискретных уравнений в свертках. Сиб. мат. ж., т. XXI (I980), № 2, II9-I27.

[2] Обратимость операторов свертки в больших областях. Мат. иссл., вып. 54 (I980), 56-66.

Krein, M.G. (Крейн, М.Г.)

[1] Интегральные уравнения на полупрямой с ядром, зависящим от разности аргументов. УМН I3 (I958), вып.5, 3-I20.

[2] О некоторых новых банаховых алгебрах и теоремах типа теорем Винера-Леви для рядов и интегралов Фурье. Мат. иссл. I:I (I966), 82-I09.

Kronecker, L.

[1] Zur Theorie der Elimination einer Variablen aus zwei algebraischen Gleichungen. Monatsber. d. Königl. Preussischen Akad. d. Wiss., Berlin, (1881), 535-600.

Krupnik, N. Ya. (Крупник, Н.Я.)

[1] Условия существования n-символа и достаточного набора n-мерных представлений банаховой алгебры. Мат. иссл., вып. 54 (1980), 84-97.

Landau, H.J. and H. Widom

[1] The eigenvalue distribution of time and frequency limiting. J. Math. Anal. Appl. 77 (1980), 469-481.

Lenard, A.

[1] Some remarks on large Toeplitz determinants. Pacific J. Math.42 (1972), 137-145.

[2] Momentum distribution in the ground state of the one-dimensional system of impenetrable Bosons. J. Math. Phys. 5 (1964), No.7, 930-943.

Linnik, I.Ju. (Линник, И.Ю.)

[1] Многомерный аналог предельной теоремы Г.Сегё. Изв. АН СССР, сер. мат., 39 (1975), № 6, 1393-1403.

Malyšev, V.A. (Малышев, В.А.)

[1] Уравнения Винера-Хопфа и их приложение в теории вероятностей. Серия "Теория вероятностей, матем. статистика, теорет. кибернетика", т. 13, Москва, 1975, 5-86.

Markus, A.S. and. I.A. Feldman (Маркус, А.С. и И.А. Фельдман)

[1] О связи между некоторыми свойствами операторной матрицы и ее определителя. Мат. иссл., вып. 54 (1980), 110-120.

[2] Об индексе операторной матрицы. Функц. анализ и его прилож. 11 (1977), № 2, 83-84.

McCoy, B. and T.T. Wu

[1] The two-dimensional Ising model. Harvard Univ. Press, 1973.

Meister, E. and F.-O. Speck

[1] Some multidimensional Wiener-Hopf equations with applications. Second symposium of frends in application of pure mathematics to mechanics, Kozubnik, Poland, 1977, 217-262.

Meyer, Ch.

[1] Das Reduktionsverfahren für singuläre Integraloperatoren mit speziellen Carleman'schen Verschiebungen. Wiss. Z. d. TH Karl-Marx-Stadt 25 (1983), Heft 3.

Michlin, S.G. (Михлин, С.Г.)

[1] Сингулярные интегральные уравнения. УМН 3 (1948), вып.3, 29-112.

Michlin, S.G. and S. Prößdorf

[1] Singuläre Integraloperatoren. Akademie-Verlag, Berlin, 1980.

Montroll, E.W., R.B. Potts, and J.C. Ward

[1] Correlations and spontaneous magnetization of the two-dimensional Ising model. J. Math. Phys. 4 (1963), No.2, 308-322.

Mullis, C. T. and R.A. Roberts

[1] Synthesis of minimum roundoff noise fixed point digital filters. IEEE-CAS 23/9 (1976), 551-562.

[2] The use of second-order information in the approximation of discrete-time linear systems. IEEE-ASSP 24/3 (1976), 226-238.

Muskhelishvili, N.I. (Мусхелишвили, Н.И.)

[1] Сингулярные интегральные уравнения. 3-е изд., Наука, Москва, 1968. (English transl.: Singular integral equations. Nordhoff, Groningen, 1953.)

Nehari, Z.

[1] On bounded bilinear forms. Ann. of Math. 65 (1957), 153-162.

Nikol'skiĭ, N.K. (Никольский, Н.К.)

[1] О пространствах и алгебрах теплицевых матриц действующих в l^p. Сиб. мат. ж., т. VII (1966), № I, 146-158.

[2] Лекции об операторе сдвига. Наука, Москва, 1980.

[3] Операторы Ганкеля и Теплица. Препринты ЛОМИ, Ленинград, 1982, I. Спектральная теория Р-I-82, II. Алгебраический подход Р-2-82, III. Последные достижения Р-5-82.

(Prof. Nikol'skiĭ communicated to us that the preprints [3] will appear as an appendix to the english translation of his book [2] by the Springer-Verlag.)

Page, L.B.

[1] Bounded and compact vectorial Hankel operators. Trans. Amer. Math. Soc. 150 (1970), 529-539.

Peller, V.V. (Пеллер, В.В.)

[1] Гладкие операторы Ганкеля и их приложения (идеалы $\mathfrak{C}_p$, классы Бесова, случайные процессы). ДАН СССР 252 (1980), № I, 43-48.

[2] Операторы Ганкеля класса $\mathfrak{C}_p$ и их приложения (рациональная аппроксимация, гауссовские процессы, проблема мажорации операторов). Мат. сборник 113 (1980), № 4, 538-581.

[3] Vectorial Hankel operators, Commutators and related operators of the Schatten-von Neumann class $\mathfrak{C}_p$. Integral Equations and Operator Theory 5 (1982), No.2, 244-272.

Peller, V.V. and S.V. Hruščev (Пеллер, В.В. и С.В. Хрущёв)

[1] Hankel operators, best approximations and stationary Gaussian processes, I - III. LOMI Preprints E-4-81, E-5-81, E-6-81, Leningrad, 1981.

[2] Операторы Ганкеля, наилучшие приближения и стационарные гауссовские процессы. УМН 37(223) (1982), № I, 53-124.

Pilidi, V.S. (Пилиди, В.С.)

[1] О многомерных бисингулярных операторах. ДАН СССР 201 (1971), № 4, 787-789.

Pincus, J.D.

[1] On the trace of commutators in the algebra of operators generated by an operator with trace class self-commutator. Unpublished, 1972.

Pomp, A.

[1] Zur Konvergenz des Reduktionsverfahrens für Wiener-Hopf'sche Gleichungen. Teil I: Ein allgemeines Operatorenschema, Preprint P-MATH-03/81, Teil II: Anwendungen auf diskrete Wiener-Hopf'sche Gleichungen und Fehlerabschätzungen. Preprint P-MATH-05/81, Akad. Wiss. DDR, Inst. f. Math., 1981.

Power, S.C.

[1] Hankel operators with PQC symbols and singular integral operators. Proc. Lond. Math. Soc., 3rd ser., 41 (1980), 45-65.

[2] Hankel operators on Hilbert space. Bull. Lond. Math. Soc. 12 (1980), 422-442.

Prößdorf, S.

[1] Einige Klassen singulärer Gleichungen, Akademie-Verlag, Berlin, 1974. (English transl.: Some classes of singular equations, North Holland, 1978.)

Prößdorf, S. and B. Silbermann

[1] Projektionsverfahren und die näherungsweise Lösung singulärer Gleichungen, Teubner-Text, Leipzig, 1977.

Rabindranathan, M.

[1] On the inversion of Toeplitz operators. J. Math. Mech. 19 (1969), 195-206.

Reed, M. and B. Simon

[1] Methods of modern mathematical physics. I: Functional analysis, Academic Press, 1972, II: Fourier analysis, Self-adjointness, Academic Press, 1975.

Richmann, D.R.

[1] A new proof of a result about Hankel operators. Integral Equations and Operator Theory 5 (1982), No.6, 892-900.

Roch, S. and B. Silbermann

[1] Das Reduktionsverfahren für Potenzen von Toeplitzoperatoren mit unstetigem Symbol. Wiss. Z. d. TH Karl-Marx-Stadt 24 (1982), Heft 3, 289-294.

[2] Toeplitz-like operators, quasicommutator ideals, numerical analysis, Math. Nachr. (to appear)

Rost, K.

[1] Invertierung einiger Klassen von Matrizen im Zusammenhang mit der näherungsweisen Lösung von Operatorgleichungen. Diss. A, TH Karl-Marx-Stadt, 1980.

Rudin, W.

[1] Real and complex analysis. McGraw-Hill, 1970.

[2] Functional analysis. McGraw-Hill, 1973.

Russakovskiǐ, E.M. (Руссаковский, Е.М.)

[1] О ганкелевых и теплицевых матрицах и безутианте, I, II. Теория функций, функц. анализ и их прилож., Харьков, 32 (I979), 77-82, 33 (I980), II9-I24.

Sachnovič, A.L. (Сахнович, А.Л.)

[1] Об одном методе обращения теплицевых матриц. Мат. иссл. 8 (I973), № 4, I80-I85.

Sarason, D.

[1] Algebras of functions on the unit circle. Bull. Amer. Math. Soc. 79 (1973), No.2, 286-299.

[2] Toeplitz operators with piecewise quasicontinuous symbols. Indiana Univ. Math. J. 26 (1977), 817-838.

Seghier, A.

[1] Inversion de la matrice de Toeplitz en plusieurs dimensions et théorème de Szegö. C.R. Acad. Sc. Paris 293, Série I (1981).

Semencul, A.A. (Семенцул, А.А.)

[1] Обращение конечных теплицевых матриц и их континуальных аналогов. В книге И.Ц.Гохберга и И.А.Фельдмана, Проекционные методы решения уравнений Винера-Хопфа, РИО АН МССР, Кишинев, I967, I40-I56.

Silbermann, B.

[1] Zur Berechnung von Toeplitz-Determinanten, die durch eine Klasse im wesentlichen beschränkter Funktionen erzeugt werden. Wiss. Z. d. TH Karl-Marx-Stadt 20 (1978), Heft 6, 683-687.

[2] Some remarks on the asymptotic behavior of Toeplitz determinants. Appl. Analysis 11 (1981), No.3, 185-197.

[3] Das asymptotische Verhalten von Toeplitzdeterminanten für einige Klassen von Erzeugerfunktionen. Abh. Akad. Wiss. DDR, Abt. Math.-Naturwiss.-Technik, Nr.2N (1981), 267-272.

[4] The strong Szegö limit theorem for a class of singular generating functions, I. Demonstratio Mathematica, vol. XIV (1981), No.3, 647-667.

[5] Ausgeartete paarige Integrodifferenzengleichungen und ein Projektionsverfahren zu ihrer Lösung. Wiss. Z. d. TH Karl-Marx-Stadt 23 (1981), Heft 4, 465-470.

[6] Lokale Theorie des Reduktionsverfahrens für Toeplitzoperatoren. Math. Nachr. 104 (1981), 137-146.

[7] Lokale Theorie des Reduktionsverfahrens für singuläre Integralgleichungen. Z. f. Analysis u. Anw. (to appear)

Silbermann, B. and K. Rost

[1] Das Reduktionsverfahren für eine Klasse ausgearteter Integrodifferenzengleichungen. Wiss. Z. d. TH Karl-Marx-Stadt 20 (1978), Heft 6, 689-691.

Simon, B.

[1] Notes on infinite determinants of Hilbert space operators. Adv. in Math. 24 (1977), 244-273.

Simonenko, I.B. (Симоненко, И.Б.)

[1] Краевая задача Римана с измеримыми коэффициентами. ДАН СССР I35 (I960), № 3, 538-54I.

[2] Краевая задача Римана для n пар функций с измеримыми коэффициентами и ее применение к исследованию сингулярных интегралов в пространствах L^p с весами. Изв. АН СССР, сер. мат., 28 (I964), № 2, 277-306.

[3] Новый общий метод исследования линейных операторных уравнений типа сингулярных интегральных операторов, I, II. Изв. АН СССР, сер. мат., 29 (I965), 567-586, 757-782.

[4] Операторы типа свертки в конусах. Мат. сборник 74(II6) (I967), 298-3I3.

[5] О многомерных дискретных свертках. Мат. иссл., вып.3, № I (I968), I08-I22.

Stein, E.M.

[1] Singular integrals and differentiability properties of functions. Princeton Univ. Press, 1971.

Szegö, G.

[1] On certain hermitian forms associated with the Fourier series of positive functions. Festkrift Marcel Riesz, Lund, 1952, 228-238.

Triebel, H.

[1] Spaces of Besov-Hardy-Sobolev type. Teubner-Text, Leipzig, 1978.

Verbickiĭ, I.E. (Вербицкий, И.Э.)

[1] О мультипликаторах в пространствах ℓ^p с весом. Мат. иссл., вып.45 (I977), 3-I6.

[2] О методе редукции для степеней теплицевых матриц. Мат. иссл., вып.47 (I978), 3-II.

[3] Проекционные методы решения сингулярных интегральных уравнений с кусочно-непрерывными коэффициентами. Мат. иссл., вып. 47 (I978), I2-24.

Verbickiĭ, I.E. and N.Ya. Krupnik (Вербицкий, И.Э. и Н.Я. Крупник)

[1] О применимости проекционного метода к дискретным уравнениям Винера-Хопфа с кусочно-непрерывным символом. Мат. иссл., вып.45 (1977), 17-28.

Vol'berg, A.L. (Вольберг, А.Л.)

[1] Two remarks concerning the theorems of S. Axler, S.-Y.A.Chang and D.Sarason. I. Operator Theory 7 (1982), No.2.

Whittaker, E.T. and G.N. Watson

[1] A course in modern analysis. 4th ed., Cambridge, 1963.

Widom, H.

[1] Inversion of Toeplitz matrices, II. Illinois J. Math. 4 (1960), 88-99.

[2] Inversion of Toeplitz matrices, III. Notices Amer. Math. Soc. 7 (1960).

[3] On the spectrum of Toeplitz operators. Pacific J. Math. 14 (1964), 365-375.

[4] Toeplitz operators on H^p. Pacific J. Math. 19 (1966), 573-582.

[5] Toeplitz determinants with singular generating functions. Amer. J. Math. 95 (1973), 333-383.

[6] Asymptotic inversion of convolution operators. Publ. Math. I.H.E.S. 44 (1975), 191-240.

[7] Perturbing Fredholm operators to obtain invertible operators. J. Functional Analysis 20 (1975), 26-31.

[8] Asymptotic behavior of block Toeplitz matrices and determinants II. Adv. in Math. 21 (1976), 1-29.

[9] Asymptotic expansions of determinants for families of trace class operators. Indiana Univ. Math. J. 27 (1978), No.3, 449-478.

[10] Families of pseudodifferential operators. In: Topics in Functional Analysis, A collection of Papers Dedicated to M.G. Krein, Academic Press, 1978, 345-395.

[11] Szegö's theorem and a complete symbolic calculus for pseudo-differential operators. In: Seminar on Singularities of solutions of Linear Partial Differential Equations, Princeton Univ. Press, 1978, 261-283.

[12] Szegö's limit theorem: the higher-dimensional matrix case. J. Functional Analysis 39 (1980), 182-198.

Wintner, A.

[1] Zur Theorie der beschränkten Bilinearformen. Math. Z. 30 (1929), 228-282.

Wu, T.T.

[1] Theory of Toeplitz determinants and spin correlations of the two-dimensional Ising model, I. Phys. Review 149 (1966), No.1, 380-402.

N O T A T I O N   I N D E X

$\mathbb{C}$ complex field

$\mathbb{D}$ open unit disk

Γ complex unit circle

$\mathbb{R}$ real field

$\mathbb{R}_{\pm}$ positive (resp. negative) real numbers

$\mathbb{Z}$ integer group

$\mathbb{Z}_{\pm}$ non-positive (resp. non-negative) integers

$\to$ strong convergence

$\rightrightarrows$ uniform convergence

SUBJECT INDEX

Bei Fragen zur Produktsicherheit wenden Sie sich bitte an:
If you have any questions regarding product safety,
please contact:

Walter de Gruyter GmbH
Genthiner Straße 13
10785 Berlin
productsafety@degruyterbrill.com